INTRODUCTORY PHYSICS WITH CALCULUS AS A SECOND LANGUAGE

THOMAS BARRETT
Ohio State University

JOHN WILEY & SONS, INC.

EXECUTIVE PUBLISHER	Kaye Pace
ACQUISITIONS EDITOR	Stuart Johnson
PRODUCTION MANAGER	Pam Kennedy
SENIOR PRODUCTION EDITOR	Sarah Wolfman-Robichaud
MARKETING MANAGER	Amanda Wygal
SENIOR DESIGNER	Kevin Murphy
CREATIVE DIRECTOR	Harry Nolan
SENIOR ILLUSTRATION EDITOR	Anna Melhorn
EDITORIAL ASSISTANTS	Alyson Rentrop, Krista Jarmas
COPYEDITOR	Constance A. Parks
PROJECT MANAGEMENT SERVICES	Shubhendu Bhattacharya/TechBooks India

This book was set in 10/12 Times Roman by GTS Companies/TechBooks India

To order books or for customer service please, call 1-800-CALL WILEY (225-5945).

ISBN-13   978-0-471-73910-4
ISBN-10   0-471-73910-3

10 9 8 7 6 5 4

PREFACE

You've read the book, you might have even heard a lecture. Now you want to know how to do the problems.

What makes languages difficult are the irregular verbs. Regular verbs are easy: I rake the yard, I raked the yard, I have raked the yard is the same as you mow the lawn, you mowed the lawn, you have mowed the lawn. Irregular verbs are not so easy: I go to the store, I went to the store, I have gone to the store or I am green with envy, I was green with envy, I have been green with envy. Irregular verbs must be memorized, because they do not fit the set pattern that regular verbs do; that is what makes them irregular.

The good news is that physics does not have irregular verbs. The solution to every physics problem fits a set pattern. Learning the pattern is much easier than learning each verb, or problem, individually.

Unfortunately, many students of physics try to learn the individual problems rather than the pattern. This is the equivalent of learning a language from a phrase book. If you want to learn physics this way, then this book is not for you. This book is about the patterns, a few techniques that can be applied to solve a variety of problems.

Fortunately, learning and following the pattern is not any more difficult than learning each problem individually.

CONTENTS

CHAPTER 1

HOW TO LEARN PHYSICS

Welcome to physics. It's not so bad—really.

The goal of physics is to learn a few things that are always true, then learn how to use them in a variety of situations. This is much easier than learning a variety of things that are occasionally true.

1.1 THE DEADLY SINS

Physics teachers have seen students do many things wrong. The following list is by no means exhaustive, nor would every physics teacher agree that these are the most common or most important.

- Expecting to solve a problem with one equation.
- (Mis) Using sample problems.
- Taking shortcuts.

EXAMPLE

Johnny is carrying groceries in from the car. He can carry two bags in his left hand and three in his right hand. How many bags can he carry in four trips?
In each trip he can carry

$$R = R_{\text{left}} + R_{\text{right}}$$

$$R = (2 \text{ bags/trip}) + (3 \text{ bags/trip})$$

$$R = 5 \text{ bags/trip}$$

In four trips he can carry

$$N = RT$$

$$N = (5 \text{ bags/trip})(4 \text{ trips})$$

$$N = 20 \text{ bags}$$

It would be possible to derive and memorize a formula for the example above: $N = (R_{\text{left}} + R_{\text{right}})T$. To learn such an equation for every possible problem would be unbearable. Was it so hard to do the problem in two steps?

EXAMPLE

It takes a janitor $\frac{1}{3}$ hour to clean a room. He has to clean 16 rooms every night. How many rooms does he clean in a 5-day workweek?

From the previous example,

$$N = RT$$

$$\text{nights } N = 5$$
$$\text{rooms } R = ?$$
$$\text{time } T = {}^1/_3$$

So

$$N = RT$$

$$5 = R\frac{1}{3}$$

$$R = 15$$

Hopefully you see that 15 is not the correct answer. The mistake is taking an equation out of context, which is very easy to do if the equation is from the solution to a different problem.

Each example physics problem is meant to teach a lesson rather than show how to solve a problem. If the purpose were to show you how to solve a problem, then you could learn only one thing from each problem. Since the real purpose is to show you how a technique works, you can apply what you learn to many problems.

EXAMPLE

How many square meters are in a square kilometer?

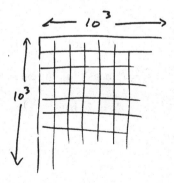

The prefix kilo- means a thousand, so 1 km^2 = 1000 m^2 = 10^3 m^2, yes? No. I've seen very good students get this wrong. They didn't think that they needed to go through the process.

$$(1\text{ km}^2)\left(\frac{10^3\text{ m}}{1\text{ km}}\right)^2 = 10^6\text{ m}^2$$

Are there shortcuts for physics problems? Sometimes, but it takes doing similar problems a few dozen times before one can recognize when the shortcut applies and how to use it.

LESSON

"Plug 'n chug" doesn't go very far.

1.2 THE CARDINAL VIRTUE

The goal of physics is to put together two (or more) simple steps to arrive at a conclusion that you couldn't have gotten otherwise.

This is so important to the study of physics that I'm going to repeat it. The solution to any physics problem is a number of steps, each of which is small enough to understand by itself. The total of all of these steps gets us to an answer. We do this all the time when driving. There is no single road that connects Nashville and New York. Instead we could take I-40 to Knoxville, I-81 to Harrisburg, and I-78 or I-80 to New York. We do not think it strange that the journey involves a few turns, nor do we expect every city to be joined to every other city by a single road. The directions to New York sound similar to the directions needed to solve a physics problem: "use conservation of momentum to find the velocity of the block of wood, then use constant acceleration or conservation of energy to determine how high it goes."

EXAMPLE

A standard incandescent bulb costs $0.25 and uses 60 watts of electricity. A compact fluorescent (CF) bulb costs $3.50 and uses 14 watts of electricity. If you pay $0.10 per kilowatt-hour (one watt for a thousand hours), how long does it take before the cost of the CF bulb is half that of the incandescent bulb?

The electrical energy used by the incandescent bulb, in kilowatt-hours, is

$$E = (60\text{ W})\left(\frac{1\text{ kW}}{1000\text{ W}}\right)(t)$$

Note that t and E are not just numbers, but have units as well (for example, $t = 24$ h). The cost of this electricity is

$$\$ = (E)(\$0.10/\text{kWh})$$

and the total cost of the incandescent bulb is

$$\$_i = \$0.25 + (60 \text{ W}) \left(\frac{1 \text{ kW}}{1000 \text{ W}} \right)(t)(\$0.10/\text{kWh})$$

Likewise the cost of the compact fluorescent bulb is

$$\$_{CF} = \$3.50 + (0.014 \text{ kW})(t)(\$0.10/\text{kWh})$$

We want the cost of the CF to be half that of the incandescent

$$\$_{CF} = \frac{1}{2}(\$_i)$$

so

$$\$3.50 + (0.014 \text{ kW})(t)(\$0.10/\text{kWh}) = \frac{1}{2}\left[\$0.25 + (0.060 \text{ kW})(t)(\$0.10/\text{kWh})\right]$$

$$\$3.375 = (0.016 \text{ kW})(t)(\$0.10/\text{kWh})$$

$$3.375 = \left(0.0016\frac{\text{kW}}{\text{kWh}}\right)t$$

$$t = \frac{3.375}{0.0016}\frac{\text{kWh}}{\text{kW}} = 2109 \text{ h}$$

This is about 5.75 hours each day for a year. The compact fluorescent may seem expensive, but if you have a light you leave on every evening, the cost of the electricity saved more than makes up for the higher initial cost!

LESSON

Expect to need more than a single step to solve a problem.

1.3 UNITS

Units are important. Really. There's a big difference between having 60 minutes to complete a test and having 60 seconds to do the test. Units will always work out if done right.

EXAMPLE

How long does it take to drive 120 miles at 60 mph?
Distance is velocity times time, so

$$t = \frac{d}{v} = \frac{120}{60} = 2$$

and times are measured in seconds, so $t = 2$ s. Hopefully you realize that this is wrong. Instead

$$t = \frac{d}{v} = \frac{120 \text{ mi}}{60 \text{ mi/h}} = \frac{120 \text{ mi}}{60} \frac{\text{h}}{\text{mi}} = 2 \text{ h}$$

EXAMPLE

A sports car accelerates from rest to 60 mph in 6.0 seconds. What is the acceleration of the sports car?

$$a = \frac{\Delta v}{\Delta t} = \frac{60 \text{ mi/h}}{6.0 \text{ s}} = 10 \frac{\text{mi}}{\text{h s}}$$

The speed of the car increases by 10 miles per hour each second.

LESSON

Keep your eye on the units.

CHAPTER 2

CONSTANT ACCELERATION

We start by trying to describe the way in which things move. We'll save the question of why they might move that way until later.

2.1 ACCELERATION AND VELOCITY

We'll start by describing the simplest possible motion. Why? Consider a complicated example: write an equation describing the position of a roller coaster as a function of time. $x(t) = x_0 + \ldots$. Uh, how 'bout we get back to the simple motion?

The simplest thing that an object might do is nothing—it could just sit there. This isn't especially exciting, but it happens sometimes. It's pretty easy to describe. In particular, if an object never moves, it's pretty easy to describe where it's been and where it's going to be.

The next least interesting thing that something could do is move with a constant velocity. **Velocity** tells us how quickly the position of the object is changing. Velocity differs from speed in that velocity has direction as well as magnitude or size. A negative velocity indicates motion in the opposite direction as a positive velocity, but the speed is positive in each case because speed doesn't have direction.

Since velocity is how fast the position is changing, we can write

$$v = \frac{\Delta x}{t} = \frac{x - x_0}{t} \qquad \text{or} \qquad x = x_0 + vt$$

Expressed in words this says that the change in the position is how fast the position is changing times how long it changes, and that the final position is the old position plus the change.

Not everyone uses the same notation, so the following are all equivalent ways of writing the same thing:

Final	Initial	Equation
x	x_0	$x = x_0 + vt$
x'	x	$x' = x + vt$
x_f	x_i	$x_f = x_i + vt$
x_{after}	x_{before}	$x_{after} = x_{before} + vt$
x_B	x_A	$x_B = x_A + vt$
change $= \Delta x$ or s		$\Delta x = vt$

I have a strong preference for the last one ($\Delta x = vt$). We'll see why in the next section.

EXAMPLE

A car travels 324 km north in 3 hours. What is its velocity (assumed to be constant)?

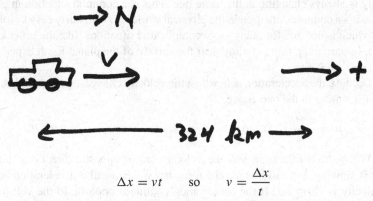

$$\Delta x = vt \qquad \text{so} \qquad v = \frac{\Delta x}{t}$$

Any time something with direction, like the displacement or the velocity, shows up in an equation, it could be positive or negative. The question is whether Δx is 324 km or −324 km. We get to choose. Coordinate systems (or axes) do not exist in nature, but are only a mathematical device, so we get to pick whether northward is positive or southward is positive. However, once we choose, we must stick with our choice throughout the problem.

For lack of anything better, and to avoid negative numbers, let's choose northward as the positive direction. Then the displacement Δx is +324 km, and

$$v = \frac{\Delta x}{t} = \frac{324 \text{ km}}{3 \text{ h}} = 108 \text{ km/h}$$

If we had chosen southward as the positive direction,

$$v = \frac{\Delta x}{t} = \frac{-324 \text{ km}}{3 \text{ h}} = -108 \text{ km/h}$$

So the car had a velocity of either 108 km/h north or −108 km/h south, which are the same thing. We have to agree on the velocity, regardless of the choice of coordinates, even though we may not call it the same thing.

LESSON

You may use any coordinate system or axes you want, but you must use the same ones for the whole problem.

If the velocity changes, we call the change in velocity the **acceleration**. If a car speeds up, it accelerates; if a car slows down, it accelerates; if a car turns, it accelerates. It may seem strange to say that a slowing car is accelerating, but acceleration is *any* change in the velocity. When a car turns, its speed may stay the same but the direction changes, so the velocity changes, so there is an acceleration.

The next least complicated thing an object can do, after having a constant velocity, is to have a constant acceleration. That is, the velocity is changing but the velocity is always changing at the same rate. This is a common situation in physics because it's a common situation in the physical world. It's also fairly easy to solve the math, which is not true for many more complicated situations (like the roller coaster above). In particular, things falling near the surface of the planet Earth experience a constant acceleration.

Because the acceleration is how fast the velocity changes, we can find an equation for it similar to the one above.

$$a = \frac{v - v_0}{t} \qquad \text{so} \qquad v - v_0 = at \qquad \text{or} \qquad v' - v = at$$

When the acceleration and the velocity are in opposite directions, then the object is slowing. We call this deceleration, but we also call it acceleration because the velocity is changing. If the acceleration continues opposite to the velocity, the object can stop and turn around. At the turnaround point, velocity is momentarily zero, but *the velocity is still changing* so the acceleration is not zero.

The units for acceleration may seem strange. The unit of position or displacement is a length, usually meters (m) but any length will do. The unit for change in displacement or change in position is also a length. The unit of velocity is change in position per time, or length per time, such as meters per second (m/s). A velocity of 3 m/s means that each second the object moves 3 meters in the positive direction. The unit of acceleration is velocity per time, such as m/s/s, or m/s^2. An acceleration of 10 m/s^2 means that each second the velocity changes by 10 m/s.

For an object falling (or rising) near the surface of the Earth, the acceleration from gravity is 9.8 m/s^2 downward. Some sources use 9.81 m/s^2, but the variation from one point on the surface of Earth to another is larger than 0.01 m/s^2. Because this value shows up so much in physics equations, it is given its own symbol: $g = 9.8\,m/s^2$. Even though the acceleration from gravity is downward, g is never negative; g is the magnitude of the acceleration of gravity.

2.2 CONSTANT ACCELERATION

A common problem in physics is to understand the motion of an object with constant acceleration. The figure shows the relationships between the relevant values. The important thing is that, if we are given any three of the five values, it is possible to draw the trapezoid and figure out all of the values. (The exception that proves the rule is when the acceleration a equals zero.)

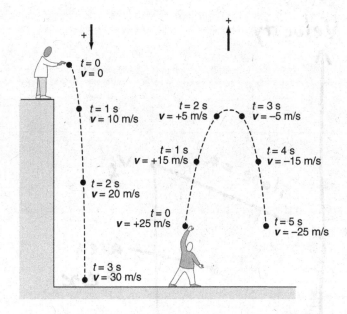

To do the math, we start with two equations: $v - v_0 = at$ and $\Delta x = v_{\text{ave}}t$. When the acceleration is constant, then the average velocity v_{ave} is equal to the average of the initial and final velocities.

$$v - v_0 = at \qquad \text{and} \qquad \Delta x = \frac{1}{2}(v_0 + v)t$$

If, for example, we knew that the initial velocity v_0 was 14 m/s, the acceleration a was 3 m/s^2, and the time t over which this happened was 12 s, then we could use the first equation to get the final velocity v and then use the second equation to get the displacement Δx.

$$v - v_0 = at$$

$$v - (14 \text{ m/s}) = (3 \text{ m/s}^2)(12 \text{ s})$$

$$v = 50 \text{ m/s}$$

$$\Delta x = \frac{1}{2}(v_0 + v)t$$

$$\Delta x = \frac{1}{2}[(14 \text{ m/s}) + (50 \text{ m/s})](12 \text{ s})$$

$$\Delta x = 384 \text{ m}$$

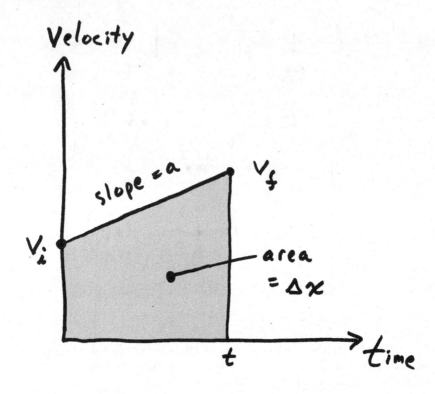

We could also do this algebraically, keeping the numbers out until the end.

$$\Delta x = \frac{1}{2}(v_0 + v)t = \frac{1}{2}[v_0 + (v_0 + at)]t = v_0 t + \frac{1}{2}at^2$$

If we worked out every possible case algebraically, we would get

Equation	Includes	Omits
$v - v_0 = at$	v_0, v, a, t	displacement Δx
$\Delta x = \frac{1}{2}(v_0 + v)t$	$\Delta x, v_0, v, t$	acceleration a
$\Delta x = v_0 t + \frac{1}{2}at^2$	$\Delta x, v_0, a, t$	final velocity v
$\Delta x = vt - \frac{1}{2}at^2$	$\Delta x, v, a, t$	initial velocity v_0
$v^2 - v_0^2 = 2a\,\Delta x$	$\Delta x, v_0, v, a$	time t

If we know *any three of the five* quantities and have a fourth that we want to find but don't care about the fifth, then pick the equation that has the three you know and the one you want but not the one you don't care about. Many physics texts use equations like $x = x_0 + v_0 t + \frac{1}{2}at^2$ instead—it's really the same equation, but then the "three of five" rule doesn't work

EXAMPLE

It takes a sports car 6.0 seconds to accelerate from rest to 60 mph. What is the acceleration of the sports car?

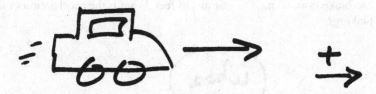

The five variables that we could know are Δx, v_0, v, a, and t. The initial velocity v_0 is zero and the time t is 6.0 seconds. The final speed is 60 mph, but the final velocity v could be +60 mph or −60 mph, since we haven't yet chosen a positive direction. Let's pick the direction the car is going at the end to be positive, so $v = +60$ mph.

$$
\begin{aligned}
\text{displacement} \quad & \Delta x = \\
\text{initial velocity} \quad & v_0 = 0 \\
\text{final velocity} \quad & v = 60 \text{ mi/h} \\
\text{acceleration} \quad & a = ? \\
\text{time} \quad & t = 6.0 \text{ s}
\end{aligned}
$$

We don't care about the distance covered by the car during the acceleration, so we choose $v - v_0 = at$:

$$v - v_0 = at$$

$$a = (v - v_0)/t$$

$$a = [(60 \text{ mi/h}) - (0)]/(6.0 \text{ s})$$

$$a = 10 \text{ mi/h/s}$$

Ten miles per hour per second is the acceleration, or every second the velocity changes by 10 mi/h. If we want this in meters and seconds,

$$a = (10 \text{ mi/h/s}) \left(\frac{1610 \text{ m}}{1 \text{ mi}} \right) \left(\frac{1 \text{ h}}{3600 \text{ s}} \right)$$

$$a = 4.5 \text{ m/s}^2$$

The acceleration is positive. This means that the acceleration is in the same direction as the final velocity of the car.

Is it necessary to memorize the five equations? Not really. If you remember any two you can figure out the one you need. This is easiest if you remember the first two: solve the first equation for the "don't care" variable and substitute it into the second equation. Many students of physics find it easier to remember all but the fourth equation, since it is rare for the initial velocity to be the "don't care" variable.

EXAMPLE

A car can brake from 60 mph to rest in 118 feet. What is the acceleration of the car while braking?

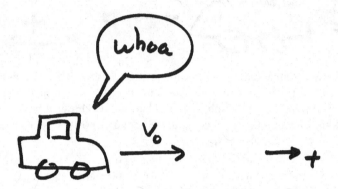

Let's choose the direction the car is initially moving as the positive direction. We make a note on the drawing to show our positive direction. Then the displacement Δx is also positive, since the car is always moving in the same direction.

$$\Delta x = +118 \text{ ft} \left(\frac{0.305 \text{ m}}{1 \text{ ft}} \right) = 36 \text{ m}$$

$$v_0 = +60 \text{ mi/h} \left(\frac{1610 \text{ m}}{1 \text{ mi}} \right) \left(\frac{1 \text{ h}}{3600 \text{ s}} \right) = 26.8 \text{ m/s}$$

$$v = 0$$
$$a = ?$$
$$t =$$

If we are determined to do the problem with metric units, then this is as good a time as any to convert them.

We don't care about the time it takes for the car to stop, so we choose $v^2 - v_0^2 = 2a \, \Delta x$.

$$v^2 - v_0^2 = 2a \, \Delta x$$

$$a = \frac{v^2 - v_0^2}{2 \, \Delta x}$$

$$a = \frac{(0)^2 - (26.8 \text{ m/s})^2}{2(36 \text{ m})}$$

$$a = -10 \text{ m/s}^2$$

The acceleration is negative. A negative acceleration does not mean that the car is slowing down—it means that the acceleration is in the opposite direction as what we chose as positive. If we had chosen the direction of the initial velocity to be negative, then

$$a = \frac{(0)^2 - (-26.8 \text{ m/s})^2}{2(-36 \text{ m})}$$

$$a = +10 \text{ m/s}^2$$

In either case the sign of a is opposite to the sign of v_0. The car is slowing down because the velocity and the acceleration are in opposite directions.

LESSON

Start a constant acceleration problem by writing down the five symbols (Δx, v_0, v, a, and t), then look for three that you know.

EXAMPLE

A boy throws a baseball straight up. It leaves his hand 2 m above the ground and reaches a peak height of 8 m above the ground. How long does it take to land on the ground?

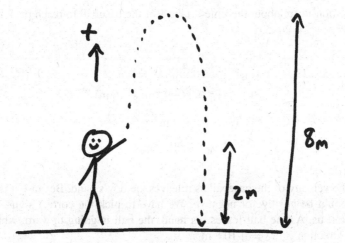

We make our list of the variables, using up as the positive direction:

$$\Delta x = -2 \text{ m}$$
$$v_0 =$$
$$v =$$
$$a = -g = -9.8 \text{ m/s}^2$$
$$t = ?$$

We need a third value in order to solve the problem.

Can we use zero for the final velocity, since, after the ball lands, it isn't moving? During its flight the baseball has an acceleration of g downward, but from the instant it hits the ground until it comes to rest (a very short time) it has a large upward acceleration, as the downward velocity changes to zero. Since this is a different acceleration, we have to use our constant acceleration equations from "the instant it leaves his hand" to "the instant it hits the ground." As it contacts the ground, the velocity of the baseball is not zero.

If all we knew was that the boy threw the baseball upward and that after an acceleration of $-g$ it hit the ground, we couldn't solve the problem. We do have one other piece of information: the peak height of the baseball. The instant of peak height is not one of our end points, so to use it we need to do another "problem." At its peak height, the baseball is neither moving up nor down, so its velocity is zero.

$$\Delta x = 6\,\text{m} \qquad\qquad\qquad \Delta x = -2\,\text{m}$$
$$v_0 = ? \qquad\qquad \rightarrow \qquad v_0 =$$
$$v = 0 \qquad\qquad\qquad\qquad v =$$
$$a = -g = -9.8\,\text{m/s}^2 \qquad\qquad a = -g = -9.8\,\text{m/s}^2$$
$$t = \qquad\qquad\qquad\qquad t = ?$$

We don't care about the time it takes for the baseball to reach peak height, so we use $v^2 - v_0^2 = 2a\,\Delta x$.

$$v^2 - v_0^2 = 2a\,\Delta x$$

$$(0)^2 - v_0^2 = 2(-9.8\,\text{m/s}^2)(6\,\text{m})$$

$$v_0 = \sqrt{2(9.8\,\text{m/s}^2)(6\,\text{m})}$$

$$v_0 = \pm 10.8\,\text{m/s}$$

This is the velocity of the baseball as it leaves the boy's hand. Because it is a square root, it could be positive or negative. We have to pick the correct sign—the math doesn't tell us. As the ball leaves his hand, the ball is going upward, which is the positive direction, so $v_0 = +10.8$ m/s.

We can now do our problem. We don't care about the final velocity (how fast the ball is going when it hits the ground), so we use $\Delta x = v_0 t + \frac{1}{2}at^2$.

$$\Delta x = v_0 t + \frac{1}{2}at^2$$

$$(-2\,\text{m}) = (+10.8\,\text{m/s})t + \frac{1}{2}(-9.8\,\text{m/s}^2)t^2$$

$$(4.9\,\text{m/s}^2)t^2 + (-10.8\,\text{m/s})t + (-2\,\text{m}) = 0$$

We use the quadratic equation:

$$t = \frac{-b \pm \sqrt{b^2 - 4ac}}{2a}$$

Note that a in the quadratic equation has nothing to do with acceleration—it's the coefficient of the squared term.

$$t = \frac{-(-10.8 \text{ m/s}) \pm \sqrt{(-10.8 \text{ m/s})^2 - 4(4.9 \text{ m/s}^2)(-2 \text{ m})}}{2(4.9 \text{ m/s}^2)}$$

$$t = \frac{(10.8 \text{ m/s}) \pm (12.5 \text{ m/s})}{(9.8 \text{ m/s}^2)}$$

$$t = -0.17 \text{ s} \quad \text{or} \quad 2.38 \text{ s}$$

Why the two answers? Look at the path of the baseball. If the constant acceleration had extended to negative times (times before it left the boy's hand), then we could travel the path backward until it "left" the ground, 0.17 seconds before reaching a height of 2 meters. We want the positive time of 2.38 seconds.

EXAMPLE

A stone dropped from a roof above takes 0.17 seconds to pass from the top of a window to the bottom of the 2.1 m high window. How far above the top of the window is the roof of the building?

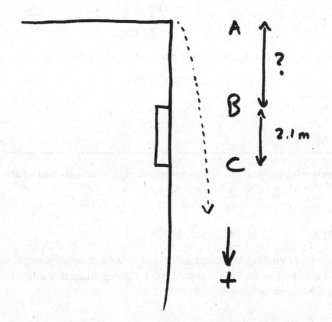

Using down as the positive direction, from the roof to the top of the window,

$$\Delta x_{AB} = ?$$
$$v_A = 0$$
$$v_B =$$
$$a = +g = +9.8 \text{ m/s}^2$$
$$t_{AB} =$$

Again we need a third value in order to solve the problem. From the top of the window to the bottom,

$$\Delta x_{BC} = 2.1 \text{ m}$$
$$v_B =$$
$$v_C =$$
$$a = +g = +9.8 \text{ m/s}^2$$
$$t_{BC} = 0.17 \text{ s}$$

We can solve for the initial velocity, which is the velocity at the top of the window. This is the final velocity for the first part, which gives us a third value.

$$\Delta x_{BC} = v_B t + \frac{1}{2} a t_{BC}^2$$

$$2.1 \text{ m} = v_B(0.17 \text{ s}) + \frac{1}{2}(+9.8 \text{ m/s}^2)(0.17 \text{ s})^2$$

$$v_B = 11.5 \text{ m/s}$$

$$v_B^2 - v_A^2 = 2a \, \Delta x_{AB}$$

$$(11.5 \text{ m/s})^2 - (0)^2 = 2(+9.8 \text{ m/s}^2) \, \Delta x_{AB}$$

$$\Delta x_{AB} = 6.75 \text{ m}$$

LESSON

When a complex problem involves more than one application of constant acceleration, keep track of which value applies to which part.

EXAMPLE

A baseball player visiting Washington, D.C., throws a baseball straight up at 44 m/s (about 100 mph). How fast will the baseball be going when it reaches the top of the Washington Monument (169 m high)?

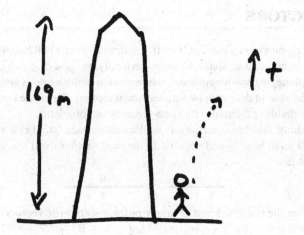

We again start with our list of five variables, taking up as positive.

$$\Delta x = +169 \text{ m}$$
$$v_0 = +44 \text{ m/s}$$
$$v = ?$$
$$a = -g = -9.8 \text{ m/s}^2$$
$$t =$$

We don't care about the time it takes for the baseball to reach a height of 169 m, so we use $v^2 - v_0^2 = 2a \, \Delta x$.

$$v^2 - v_0^2 = 2a \, \Delta x$$

$$(v)^2 - (44 \text{ m/s})^2 = 2(-9.8 \text{ m/s}^2)(169 \text{ m})$$

$$v = \sqrt{(44 \text{ m/s})^2 - 2(9.8 \text{ m/s}^2)(169 \text{ m})}$$

$$v = \sqrt{-1376 \text{ m}^2/\text{s}^2} = \sqrt{-1376} \text{ m/s}$$

How do we take the square root of a negative number? We don't! Either we made a mistake or the answer doesn't exist. Could even a good player throw a baseball over 500 ft high? Since the ball never gets that high, it is meaningless to ask how fast it is going when it gets that high, and the math reflects this.

LESSON

Minus signs are important.

2.3 VECTORS

Not everything moves in a straight line. If you throw a tennis ball horizontally, it does not move only horizontally but also drops vertically because of gravity. Such motion (in a single plane) is two-dimensional, while motion along a line is one-dimensional. We need to be able to deal with two-dimensional motion. We tackle two-dimensional problems by dividing them into two one-dimensional problems.

Something that has direction as well as magnitude (size) is a **vector**. In this section we'll learn how to add vectors. In the next section we'll use vectors to deal with tennis balls.

$$\overset{\textstyle A}{\underset{\textstyle 3}{\longrightarrow}} \qquad \overset{\textstyle B}{\underset{\textstyle 4}{\longrightarrow}}$$

Consider the vectors A and B (many physics texts write vectors in bold so that the reader will know that it's a vector). What is $A + B$? Not 7, but "7 to the right." When we add two vectors we get another vector, which also has direction.

$$\overset{\textstyle C}{\underset{\textstyle 4}{\longleftarrow}}$$

What is $A + C$? Either "1 to the left" or "−1 to the right." Two people might choose different coordinate systems and would give the answer differently, but they would draw the same vector as the answer. Adding parallel or antiparallel vectors is simple (A and C are antiparallel).

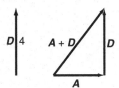

What is $A + D$? Because they are perpendicular to one another, they form a right triangle. We can use the pythagorean theorem to find the result: $\sqrt{3^2 + 4^2} = 5$. We can use trigonometry to find the lower-left angle:

$$\tan \phi = \frac{\text{opposite}}{\text{adjacent}} = \frac{4}{3}$$

$$\phi = \arctan\left(\frac{4}{3}\right) = 53°$$

What is $A + E$? Adding vectors that are parallel or perpendicular are straightforward, but A and E are neither. What we do is divide the vector E into two pieces, E_x and E_y. Since A and E_x are parallel they are simple to add. E_y is perpendicular to this so we can do that. We use trigonometry to divide E into E_x and E_y.

$$\cos 55° = \frac{\text{adjacent}}{\text{hypotenuse}} = \frac{E_x}{E}$$

$$E_x = E \cos 55° = (4)\cos 55° = 2.29$$

$$\sin 55° = \frac{\text{opposite}}{\text{hypotenuse}} = \frac{E_y}{E}$$

$$E_y = E \sin 55° = (4)\sin 55° = 3.28$$

$$A + E = A + E_x + E_y = (A + E_x) + E_y$$

$$A + E_x = 3 + 2.29 = 5.29$$

$$\sqrt{5.29^2 + 3.28^2} = 6.22$$

This is the procedure whenever we add vectors:

- Choose a set of coordinate axes (they don't have to be horizontal and vertical, but they do have to be perpendicular to each other).
- Divide each vector into pieces parallel to the axes, called components.
- Add the parallel components.
- This should leave us two pieces, which are perpendicular to each other, so we add them using the Pythagorean theorem.

The x component is not always cosine and the y component is not always sine. Instead, the adjacent side of the triangle is cosine and the opposite side is sine. Also, keep in mind that components can be negative.

EXAMPLE

Add the vectors F and G.

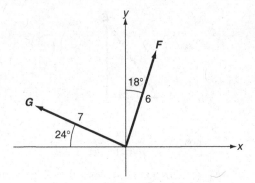

We divide **F** into two pieces, one parallel to the x-axis and the other parallel to the y-axis. The x piece is opposite to the 18° angle, so it's sine.

$$F_x = 6\sin 18° = 1.85$$

The y piece is adjacent to the 18° angle, so it's cosine.

$$F_y = 6\cos 18° = 5.71$$

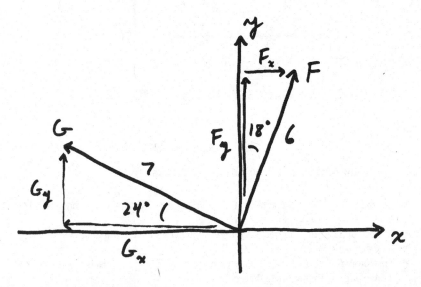

We likewise divide **G** into two pieces, one parallel to the x-axis and the other parallel to the y-axis. The x piece is adjacent to the 24° angle, so it's cosine, but it goes in the opposite direction to the axis so it's negative.

$$G_x = -7\cos 24° = -6.39$$

The y piece is opposite to the 24° angle, so it's sine.

$$G_y = 7\sin 24° = 2.85$$

A component is negative when it goes in the opposite direction as the axis.

We add the x components and add the y components:

$$(\mathbf{F} + \mathbf{G})_x = F_x + G_x = (1.85) + (-6.39) = -4.54$$

$$(\mathbf{F} + \mathbf{G})_y = F_y + G_y = (5.71) + (2.85) = 8.56$$

The resultant vector goes 4.54 to the left (opposite the x-axis) and 8.56 up. The magnitude of the resultant vector is

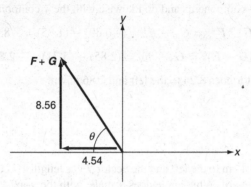

$$|\mathbf{F} + \mathbf{G}| = \sqrt{(-4.54)^2 + (8.56)^2} = 9.69$$

The angle θ that the resultant makes with the $-x$-axis is

$$\theta = \arctan\left(\frac{\text{opposite}}{\text{adjacent}}\right) = \arctan\left(\frac{8.56}{4.54}\right) = 62°$$

EXAMPLE

Find the vector difference $\mathbf{G} - \mathbf{F}$.

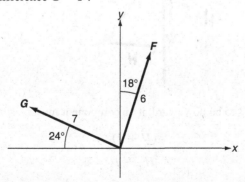

Subtracting vectors graphically is challenging—even smart people often get it backward. We want a vector that when added to $\mathbf{F}$ gives $\mathbf{G}$. The vector $\mathbf{G} - \mathbf{F}$ points to the left and down. It's usually easier to add the negative vector, $\mathbf{G} + (-\mathbf{F})$.

Like before we find the components.

$$F_x = 6 \sin 18° = 1.85$$
$$F_y = 6 \cos 18° = 5.71$$
$$G_x = -7 \cos 24° = -6.39$$
$$G_y = 7 \sin 24° = 2.85$$

We subtract the x components and do likewise with the y components.

$$(G - F)_x = G_x - F_x = (-6.39) - (1.85) = -8.24$$
$$(G - F)_y = G_y - F_y = (2.85) - (5.71) = -2.86$$

The resultant vector goes 8.24 to the left and 2.86 down.

EXAMPLE

The vector H is 3.1 cm to the left and the vector J has length 4.7 cm. The sum $H + J$ is vertically downward. What angle does J make with the y-axis?

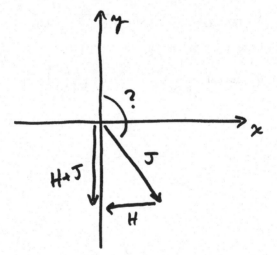

In order for $H + J$ to be downward, its x component must be zero,

$$(H + J)_x = 0$$
$$H_x + J_x = 0$$

Since H is horizontal it only has an x component, and $H_x = -|H| = -3.1$ cm. Vector J has a magnitude $|J| = J = 4.7$ cm and an x component $J_x = -H_x = +3.1$ cm. Because the x component is opposite to the angle ϕ, the x component is the sine.

$$J_x = J \sin \phi$$

$$+3.1 \text{ cm} = 4.7 \text{ cm} \ \sin \phi$$

$$\sin \phi = \frac{+3.1 \text{ cm}}{4.7 \text{ cm}} = \frac{3.1}{4.7} = 0.6596$$

$$\phi = \sin^{-1}(0.6596)$$

This step can be tricky. There are two angles in the circle ($0°$ to $360°$ or $-180°$ to $180°$) that have a sin of 0.6596, $41.3°$ and $138.7°$. Whenever you find an inverse trig function your calculator will tell you only one of the two values, so you must figure out the other and determine which is needed. If we used $41.3°$, then the vertical component of J would be upward instead of downward, so we need the other value, $138.7°$.

2.4 PROJECTILES

Sometimes an object flies through the air horizontally as well as vertically. The acceleration is still vertically downward while the velocity has a horizontal component. Because these are neither parallel nor antiparallel, we can't put the magnitudes into our constant acceleration equations and solve. Instead we break each vector into components and solve each component separately (these components don't have to be horizontal and vertical, but they need to be perpendicular).

The vertical direction is often called the y direction, though you could use anything (x, z, or *bert*, for example). The acceleration in the vertical direction is constant: g downward. Therefore we can use all of the constant acceleration equations. In order to keep track of when we're doing y components and when we're doing x components, we use little x and y subscripts, so

$$\Delta x = v_0 t + \frac{1}{2} a t^2 \qquad \text{becomes} \qquad \Delta x = v_{x0} t + \frac{1}{2} a_x t^2$$

for the horizontal direction and

$$v^2 - v_0^2 = 2a \ \Delta x \qquad \text{becomes} \qquad v_y^2 - v_{y0}^2 = 2a_y \ \Delta y$$

for the vertical direction. These aren't new equations, just new names for the values in the same equations. Δx is the "horizontal displacement," v_{y0} is the "vertical initial velocity," and t_y is the "vertical time" :-). Since time isn't a vector it doesn't have components. The time the projectile is traveling horizontally is the same as the time it is traveling vertically. This is the connection between the horizontal and vertical problems: the time t is the same.

EXAMPLE

Joe throws a ball at a brick wall 23 m away. He throws the ball at 19 m/s at $25°$ above horizontal. If the ball leaves his hand 1.7 m above ground, how far above ground does the ball hit the wall?

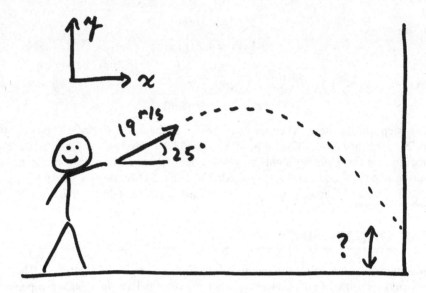

We divide the problem into horizontal x and vertical y parts. The acceleration is constant along each axis.

$$\Delta x = +23 \text{ m} \qquad\qquad\qquad \Delta y = ?$$
$$v_{x0} = +(19 \text{ m/s}) \cos 25° \qquad\qquad v_{y0} = +(19 \text{ m/s}) \sin 25°$$
$$v_x = \qquad\qquad\qquad\qquad\qquad v_y =$$
$$a_x = 0 \qquad\qquad\qquad\qquad\qquad a_y = -g$$
$$t = \qquad\qquad \longleftrightarrow \qquad\qquad t =$$

We are trying to find the vertical displacement y. Unfortunately, we only know two of the five quantities in the vertical direction, so we can't solve for y. We do know three of the five in the horizontal direction, so we can solve for v_x and t. The time t is the same for the horizontal and vertical directions, so if we solve the horizontal problem for t then we can use that as our third value in the vertical problem.

$$\Delta x = v_{x0}t + \frac{1}{2}a_x t^2$$

$$(23 \text{ m}) = [(19 \text{ m/s}) \cos 25°]\, t + \frac{1}{2}(0)t^2$$

$$t = \frac{(23 \text{ m})}{(19 \text{ m/s}) \cos 25°}$$

$$t = 1.34 \text{ s}$$

Now we can solve the vertical problem for y.

$$\Delta x = v_0 t + \frac{1}{2}at^2 \rightarrow \Delta y = v_{y0}t + \frac{1}{2}a_y t^2$$

$$\Delta y = [(19 \text{ m/s}) \sin 25°] (1.34 \text{ s}) + \frac{1}{2}(-9.8 \text{ m/s}^2)(1.34 \text{ s})^2$$

$$\Delta y = 2.0 \text{ m}$$

The ball hits the wall 2.0 m above where it left Joe's hand, or 3.7 m above the ground. If Δy had been -1.7 m or even more negative, then the ball would have hit the ground before it hit the wall.

EXAMPLE

A cannon is at the top of a 80 m high cliff overlooking flat ground. It fires a cannonball at 40 m/s at an angle of 37° above horizontal. How far from the base of the cliff does the cannonball land?

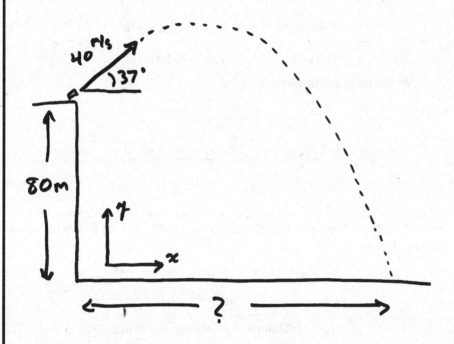

We divide the problem into horizontal x and vertical y parts. The acceleration is constant along each axis.

$\Delta x = ?$	$\Delta y = -80 \text{ m}$
$v_{x0} = +(40 \text{ m/s}) \cos 37°$	$v_{y0} = +(40 \text{ m/s}) \sin 37°$
$v_x =$	$v_y =$
$a_x = 0$	$a_y = -g$
$t =$	$t =$

We are trying to find the horizontal displacement x. Since the horizontal acceleration a_x is zero, $v_x = v_{x0}$. Unfortunately, this is the exception to the "three of five" rule—when the acceleration is zero. But if the acceleration is zero, then the velocity doesn't change, so we can find the horizontal displacement with a constant horizontal velocity:

$$\Delta x = v_{x0}t$$

To find x we need to know t, which we don't know.

Let's look at the vertical part of the problem. We know three of the five, so we can solve for the other two. In particular, we can solve for the time t. This is the same time as in the horizontal part—the cannon ball stops moving horizontally when it lands. We solve the vertical part for t, then put it in the horizontal part.

In the vertical part, we don't care about v_y so

$$\Delta x = v_0 t + \frac{1}{2}at^2 \rightarrow \Delta y = v_{y0}t + \frac{1}{2}a_yt^2$$

$$(-80 \text{ m}) = (+24 \text{ m/s})t + \frac{1}{2}(-9.8 \text{ m/s}^2)t^2$$

$$(4.9 \text{ m/s}^2)t^2 + (-24 \text{ m/s})t + (-80 \text{ m}) = 0$$

We use the quadratic equation.

$$t = \frac{-b \pm \sqrt{b^2 - 4ac}}{2a}$$

$$t = \frac{-(-24 \text{ m/s}) \pm \sqrt{(-24 \text{ m/s})^2 - 4(4.9 \text{ m/s}^2)(-80 \text{ m})}}{2(4.9 \text{ m/s}^2)}$$

$$t = -2.28 \text{ s} \qquad \text{or} \qquad 7.17 \text{ s}$$

For those who detest quadratics, there is a way to avoid it. If we knew the final vertical velocity v_y, we would know four of the five and we could pick the equation we liked to solve for t.

$$v^2 - v_0^2 = 2a \, \Delta x \rightarrow (v_y)^2 - v_{y0}^2 = 2a_y \, \Delta y$$

$$(v_y)^2 = v_{y0}^2 + 2a_y \, \Delta y$$

$$v_y = \sqrt{(24 \text{ m/s})^2 + 2(-9.8 \text{ m/s}^2)(-80 \text{ m})}$$

$$v_y = \pm 46.3 \text{ m/s}$$

Since the cannonball is going down, and since we chose up as positive, $v_y = -46.3$ m/s. Now we solve for the time t.

$$v - v_0 = at \rightarrow v_y - v_{y0} = a_yt$$

$$t = \frac{v_y - v_{y0}}{a_y}$$

$$t = \frac{(-46.3 \text{ m/s}) - (24 \text{ m/s})}{(-9.8 \text{ m/s}^2)}$$

$$t = 7.17 \text{ s}$$

We're really doing the same math—compare the two square roots—and choosing the sign for v_y is the same as choosing the sign in the quadratic equation. But often it's easier to do something in two easy steps rather than one difficult step (which is the theme of this book).

Now we do the horizontal part of the problem.

$$x = v_{x0}t$$

$$x = (32 \text{ m/s})(7.17 \text{ s})$$

$$x = 230 \text{ m}$$

LESSON

Divide two-dimensional problems into a pair of one-dimensional problems.

EXAMPLE

A cannon is at the top of a 80 m high cliff overlooking ground that tilts down at 15°. It fires a cannonball at 40 m/s at an angle of 37° above horizontal. How far from the base of the cliff does the cannonball land?

If we use horizontal and vertical as our axes, then

$\Delta x = ?$ $\Delta y = (-80 \text{ m} - x \tan 15°)$
$v_{x0} = +(40 \text{ m/s}) \cos 37°$ $v_{y0} = +(40 \text{ m/s}) \sin 37°$
$v_x =$ $v_y =$
$a_x = 0$ $a_y = -g$
$t =$ $\longleftrightarrow$ $t =$

We could solve this, but it would be ugly, because of the connection between the x and y displacements.

If instead we make our x-axis parallel to the slanted ground, and y perpendicular to x, then

$\Delta x = d + (80 \text{ m}) \sin 15°$ $\Delta y = -(80 \text{ m}) \cos 15°$
$v_{x0} = +(40 \text{ m/s}) \cos 52°$ $v_{y0} = +(40 \text{ m/s}) \sin 52°$
$v_x =$ $v_y =$
$a_x = +g \sin 15°$ $a_y = -g \cos 15°$
$t =$ $\longleftrightarrow$ $t =$

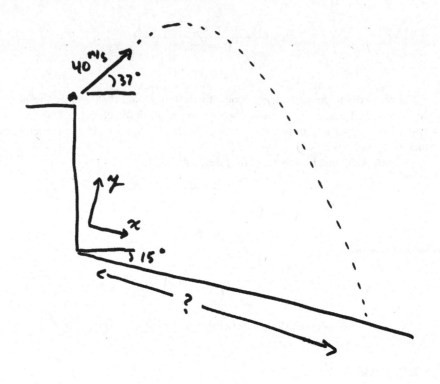

where d is the distance from the bottom of the cliff to the landing point. We can solve this the same way we did the previous problem.

$$\Delta x = v_0 t + \frac{1}{2}at^2 \rightarrow \Delta y = v_{y0}t + \frac{1}{2}a_y t^2$$

$$(-77.3\text{ m}) = (+31.5\text{ m/s})t + \frac{1}{2}(-9.47\text{ m/s}^2)t^2$$

$$(4.73\text{ m/s}^2)t^2 + (-31.5\text{ m/s})t + (-77.3\text{ m}) = 0$$

$$t = \frac{-(-31.5\text{ m/s}) \pm \sqrt{(-31.5\text{ m/s})^2 - 4(4.73\text{ m/s}^2)(-77.3\text{ m})}}{2(4.73\text{ m/s}^2)}$$

$$t = -1.91\text{ s   or   } 8.57\text{ s}$$

$$\Delta x = v_{x0}t + \frac{1}{2}a_x t^2$$

$$[d + (80\text{ m})\sin 15°] = (24.6\text{ m/s})(8.57\text{ s}) + \frac{1}{2}(2.54\text{ m/s}^2)(8.57\text{ s})^2$$

$$d = 283\text{ m}$$

(The purpose of this example is to show axes that aren't horizontal and vertical.)

EXAMPLE

Three cannons each fire a cannonball at a target. The paths of the cannonballs are shown. Rank the cannonballs by initial speed, from highest to lowest.

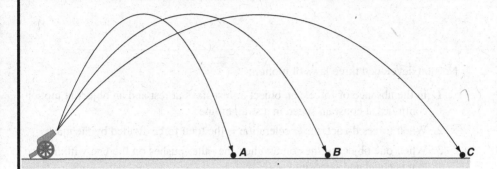

Look first at the vertical part of each problem, just to the peak height. All three have the same vertical displacement, the same vertical acceleration $(-g)$, and the same final vertical velocity (zero at the peak). Since we know three of the five, we can apply our constant acceleration technique. Therefore, they must have the same initial vertical velocity, the same time to reach the peak and the same time to land. Vertically, the three are identical.

Horizontally, C goes farther than A in the same time, so the horizontal velocity of C must be greater than A. Since the vertical velocities of A and C are the same, C must have a greater initial speed than A.

$$v_C > v_B > v_A$$

CHAPTER SUMMARY

- Solve constant acceleration problems using the "three of five" rule.

Equation	Omits
$v - v_0 = at$	displacement Δx
$\Delta x = \frac{1}{2}(v_0 + v)t$	acceleration a
$\Delta x = v_0 t + \frac{1}{2}at^2$	final velocity v
$\Delta x = vt - \frac{1}{2}at^2$	initial velocity v_0
$v^2 - v_0^2 = 2a\,\Delta x$	time t

- Divide two-dimensional problems into two problems using the perpendicular axes of your choice.
- When finding components, cosine is adjacent and sine is opposite.

CHAPTER 3

FORCES

Newton developed three laws of motion:

1. In the absence of forces, an object at rest stays at rest and an object in motion continues at constant speed in a straight line.
2. When forces do act, the acceleration is the total force divided by the mass.
3. When one object pushes on another, the other pushes on the one with a force that is equal and opposite.

The first law says that if there are no forces, then the motion doesn't change. Not only did the ancient Greeks not figure this out, neither did anyone else for 2000 years. That's not to say that this is difficult, but it might not be intuitive. The problem is that there is nowhere on Earth where there are no forces.

Newton's second law is $F = ma$. Newton says that if you add all of the forces together to get the total force or net force, then that total force is equal to the mass times the acceleration. Mass is a measure of how much stuff there is. Mass is measured in kilograms (1 kg = 1000 grams). The force is measured in newtons:

$$(1 \text{ kg})(1 \text{ m/s}^2) = 1 \text{ kg m/s}^2 = 1 \text{ N}$$

Newton's third law is sometimes said, "For every action there is an equal and opposite reaction." That statement of the third law is confusing. A more accurate restatement is, "If $\mathcal{A}$ pushes $\mathcal{B}$, then $\mathcal{B}$ pushes $\mathcal{A}$."

The first step in any force problem is to construct a free body diagram. Once we have the free body diagram, we need to choose a set of coordinate axes. We're going to be adding vectors, and we need to be able to specify our directions in a consistent manner. Then we'll use the diagram to write our equations. Finally we'll solve the equations to get our answer.

3.1 FREE BODY DIAGRAMS

In this chapter we'll use the second law, $F = ma$, over and over and over again. To find the total (or net) force, we need to add all of the forces acting on an object (be it a box or a duck). To this end we start by making a free body diagram. **The purpose of the free body diagram is to make a complete list of the forces on an object and get the directions correct.**

On a recent quiz, on a multiple choice problem that involved only two forces, students who tried to draw a free body diagram got the problem correct about 75% of the time, while those who did not scored slightly better than guessing.

LESSON

You cannot expect to do a problem with forces correctly without a free body diagram.

When we speak of a free body diagram, the word free means unattached, like a bachelor, rather than inexpensive. We draw a diagram for a single object. The diagram shows all of the forces acting on that object.

There are many kinds of forces. Anything that pushes or pulls on an object is a force. Ropes, springs, gravity, the floor—even the air can exert a force on an object. Forces can be divided into two types (anything can be divided into two types, but it does help to think of forces that way). Most forces are exerted by something that touches the object. Ropes, springs, the floor, and the air all fit this category. Other forces can push or pull on an object without touching them. The important "long-distance" forces are gravity and electricity. We will deal with electricity later.

EXAMPLE

A ballplayer throws a baseball into the air, as shown. Draw a free body diagram at each labeled position.

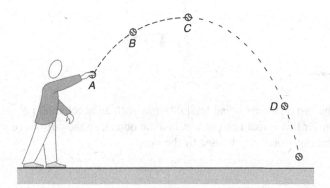

As he throws the ball, at point *A*, he is in contact with it pushing it up and to the right. Nothing else is in contact with the ball, so any other forces must be long-distance forces. Since we haven't learned about electricity yet, the only other possible force is gravity. Gravity pulls the ball down with a force of magnitude mg. (Some people write W and put in the magnitude mg in later.)

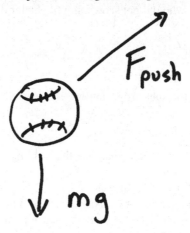

After the ball leaves his hand, there is nothing in contact with it. The only force acting on the ball is gravity. This is true at points $B - D$ and at all points along the ball's flight, until the moment that it touches the ground.

In the last chapter we had lots of things with an acceleration of g downward. The reason for this is that nothing touched the object, so the only force was gravity. The gravitational force mg divided by the mass m is g.

LESSON

The force of gravity is always down, even if the object is moving up or not moving at all.

One force that we will systematically ignore is air resistance. As an object moves through the air, it pushes the air forward while the air pushes it backward. The reason that a feather falls slower than a brick is that the effect of the air resistance on the feather is greater (which is not the same as saying that the force of the air resistance is greater).

Why do we ignore the air resistance? For many objects, the force of air resistance is much smaller than any of the other forces. But the real reason is that air resistance is too hard to calculate. The force of air resistance changes as the speed of an object through the air changes. If the force changes, the acceleration changes, and all of the stuff that we learned earlier is replaced by something much more difficult. Without air resistance, an entry-level physics student can solve the problem in a few minutes. With air resistance, an advanced physicist can't solve the problem algebraically. Since we like solvable problems, we'll ignore air resistance.

EXAMPLE

A child swings back and forth on a swing set. Draw a free body diagram for the child and swing.

The child is near Earth, so she experiences gravity. Gravity points straight down, and has a magnitude mg. Gravity and electricity are the only noncontact forces, and we don't have electricity yet.

What contact forces are there? There is a chain connected to the swing, so it could exert a force. Chains (and strings, ropes, and steel cables) exert a force in the direction that they pull (try tying a rope around a refrigerator and pushing the refrigerator with the rope). The magnitude of this force is equal to the tension in the chain.

Do we know the tension in the chain holding the swing? No, we are not told, and it is not immediately obvious. It is not true that the tension is equal to the weight of the child (in this case it is equal to $mg \sin \theta$, where θ is the angle the chain makes with the vertical). Since we don't know the value of the tension force, only its direction, we give the magnitude a symbol or name. Typical symbols used for tensions are T and F_T.

Notice that these forces could not possibly add to give zero. The velocity of the child is changing, even if the velocity is zero at the time, so the acceleration of the child is not zero, so the total force is not zero.

Whenever you encounter a force with an unknown magnitude, give the magnitude a symbol or name so that we can refer to it, talk about it, and put it in an equation. Maybe we'll be able to solve for it later; maybe it'll cancel and we won't have to. The symbol is the magnitude of the force and is a positive value, even if the force is pointing in the negative direction. We can use almost anything as this symbol except symbols already used in the problem (such as m and g), but it's helpful to choose a symbol that other people will associate with the force in question. Using a would be a poor choice because someone might think it was related to the acceleration, for example. Besides, we might want to find the acceleration later.

EXAMPLE

A rope and pulley lift a box with an acceleration of 0.4 m/s^2 upward. Draw a free body diagram for the box.

The box is near Earth, so there is a gravity force mg going toward Earth. The only object in contact with the box is the rope, which pulls the box upward. We don't know how strong the tension force is, so we give it a symbol, like T.

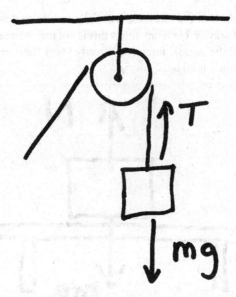

Since the box is accelerating upward, the total force points upward. If we add the two forces acting on the box, we must get upward, so the tension is greater than gravity ($T > mg$). Eventually we'll want to find out just how strong the tension is.

LESSON

Strings exert a tension force. This force is always in the direction that the string pulls. Unless we are told how big this force is, we don't know.

When an object is in contact with a surface, the surface can push on the object. The reason that you do not fall into the earth is that the ground (or floor) pushes up on you. This force is always perpendicularly outward from the surface and just hard enough to keep you from falling into the surface. We could call this the "perpendicularly out of the surface" force, but that would get cumbersome. Math people have a word that means "perpendicularly out of the surface" and so we borrow the word from them: normal. The **normal force** is not the usual or typical force, but the "perpendicularly out of the surface" force.

EXAMPLE

A box sits on a table. Draw a free body diagram for the box.

The box is on the table, which is on Earth, so there is a gravitational force mg going toward Earth. The only object in contact with the box is the table, which could push on the box. The normal force must be perpendicular to the surface, or upward in this case.

We do not know how big this force will be. It may seem that the normal force is equal to the force of gravity, but many times this is not true. Instead we pick a symbol for the magnitude of the normal force (we already know the direction) For a normal force, the symbols often used are N or F_N.

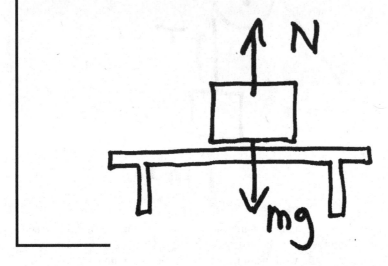

EXAMPLE

A girl stands in an elevator as it moves from the first floor to the fifth floor. Draw a free body diagram for the girl.

Since the girl is near Earth, gravity is acting on her. Gravity pulls down with a force of $W = mg$.

She is in contact with the floor of the elevator. This surface could exert a normal force. Any normal force from the floor would be upward, perpendicular to the floor. We don't know how strong this force would be. When the elevator first starts, she accelerates upward, so the normal force must be stronger than gravity. When the elevator stops at the fifth floor, she accelerates downward, so the normal force must be weaker than gravity.

LESSON

Whenever an unknown force appears in the free body diagram, give it a unique symbol. This symbol is the magnitude of the force.

EXAMPLE

A rope pulls a box across a frictionless floor. The rope is 29° above horizontal. Draw a free body diagram for the box.

There is a gravity force mg going toward Earth. The box is being pulled by a rope, so we add a tension force. The box is in contact with the floor, so we add a normal force out of the floor. We don't know the size of either the tension or the normal force, so we give them symbols.

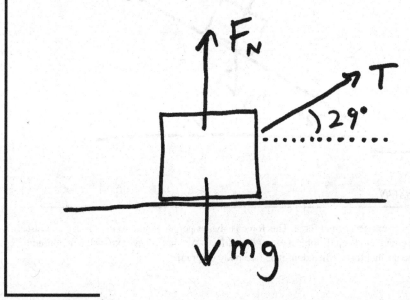

EXAMPLE

A box is placed on a 32° frictionless ramp. Draw a free body diagram for the box.

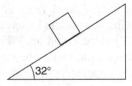

Once again there is a gravity force mg going down, toward Earth. The box is in contact with the ramp, so we add a normal force perpendicularly out of the ramp. We don't know the magnitude of the normal force so we give it the symbol N.

Part of the free body diagram is to get the directions of the forces. Draw a line along the normal force to the ground. This line forms a triangle along with the ramp and the ground since the normal force is perpendicular to the ramp, angle b and $32°$ add to $90°$. There is another triangle formed by the angles a, b, and c. Since gravity is vertical, angle c is a right angle, and angle b and angle a add to $90°$. Therefore angle a is $32°$, and angle d is also $32°$.

It is not possible for these forces to add to zero. Therefore the acceleration of the box will not be zero. The box would not stay put on a slippery (frictionless) surface.

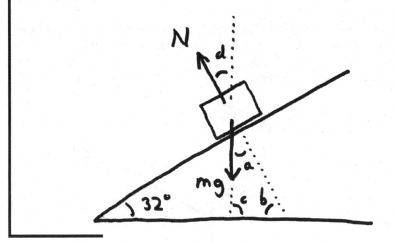

LESSON

Surfaces exert a normal force. This force is always perpendicular to the surface and just hard enough to keep the object out of the surface. We almost never know the magnitude of the normal force when doing the free body diagram.

EXAMPLE

A man on a motorcycle rides around a track in a vertical loop. Draw a free body diagram for the rider and motorcycle (together) at the top of the loop.

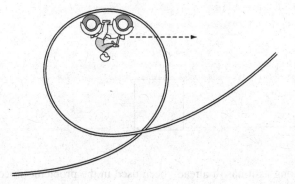

There is a gravity force *mg* going straight down, toward Earth. There is only one thing in contact with the motorcyclist: the track. The track exerts a normal force F_N perpendicularly out from the track.

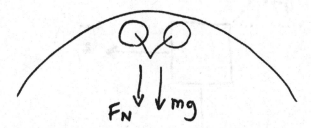

At the top of the loop both forces point down. The motorcycle accelerates downward—it would not stay there (don't try it).

LESSON

It is not a given that the forces add to zero.

EXAMPLE

A man holds a box off of the floor by pulling on a rope, as shown. Draw a free body diagram for the box.

There is a gravity force of *mg* pointing downward. The box is not in contact with any surfaces, so there are no normal forces. There are two ropes pulling on the box, so we add two tension forces.

Is it obvious that the tensions in the two ropes must be equal? It is possible that they are, but it is not clear that they must be. If the magnitudes aren't necessarily the same, then we can't use the same symbol for the forces. We can use any symbol

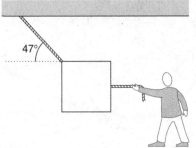

we want as long as it hasn't already been used in the problem. Since they are both tensions, we can use T_1 and T_2.

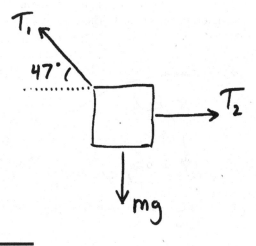

EXAMPLE

A box hangs from a rope. Tied to the bottom of the box is another rope, and a second box is tied to this rope. Ignore the mass of the ropes and draw free body diagrams for the two boxes.

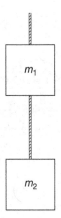

Each box has mass, so each box has gravity pulling on it. We aren't told that the masses of the boxes are the same, so we need different symbols for the two masses, like m_1 and m_2. The weights of the boxes are then m_1g and m_2g.

The top rope pulls up on the top box and the bottom rope pulls down on the top box. The forces do not have to be the same, so we pick different symbols for the magnitude of these forces, T_1 and T_2.

The bottom rope pulls up on the bottom box. Is this force necessarily the same as either of the other tensions? The tension is the same everywhere along the bottom rope, so the force pulling up on the bottom box is T_2.

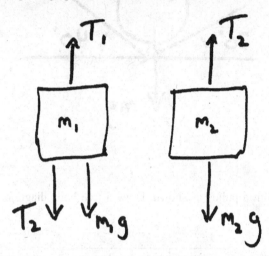

Why is the tension the same everywhere along the rope? If the rope is massless (purchased at the theoretical physics store) then the mass times the acceleration of the rope must be zero no matter what the acceleration is, so the total force on the rope is zero. The force pulling the rope up is the same as the force pulling the rope down.

EXAMPLE

A can rests between two boards, which are 30° and 40° from horizontal. Draw a free body diagram for the can.

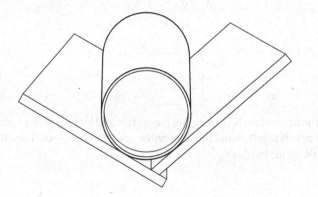

There is a gravity force of mg pointing downward. The can is in contact with two surfaces, each of which exerts a normal force on the can. Do the two normal forces have the same magnitude? There is no reason that they must. Since we don't know that they have the same strength we give them different symbols: N_1 and N_2.

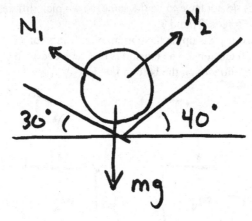

EXAMPLE

A box hangs from a pulley as shown. Draw a free body diagram for the box and bottom pulley.

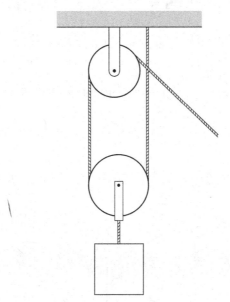

We can draw a free body diagram for the box and bottom pulley together because they are attached together and move together, as a single object. They have mass, so gravity pulls them down.

The only thing in contact with the box and pulley is the rope. The tension is the same everywhere along the rope, so we label the ropes going up from the pulley with the same symbol T.

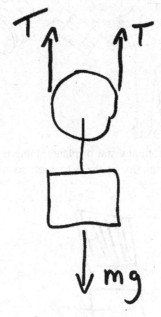

There are similar cases where the two tensions are not the same. They involve a pulley with non-zero mass and non-zero acceleration.

LESSON

Forces may be given the same symbol only if we know that they must have the same magnitude.

Newton's third law is, "If $\mathcal{A}$ pushes $\mathcal{B}$, then $\mathcal{B}$ pushes $\mathcal{A}$." For example, a book is placed on a table. Earth pulls down on the book with a gravitational force mg. Newton's third law says that the book pulls up on Earth with a force mg, of the same magnitude but in the opposite direction. It is true that the table pushes up on the book with a normal force, and it is also true that sometimes this force is equal and opposite to gravity (if there are no other forces and $a = 0$). Even if the normal force is equal and opposite to gravity it is not the force Newton was talking about in his third law.

EXAMPLE

An astronaut in outer space pushes on a 600 kg satellite with a force of 100 N. Draw a free body diagram for the satellite.

In outer space, far from any star or planet, there is no gravity force. The only thing in contact with the satellite is the astronaut, so we include the force of him pushing on the satellite.

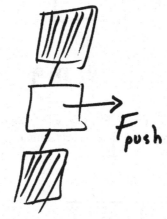

Where is the equal and opposite force from Newton's third law? If the astronaut pushes on the satellite, then the satellite pushes on the astronaut. This is the free body diagram of the astronaut.

EXAMPLE

Two boxes rest on a frictionless floor. Joe pushes the left box to the right. Draw a free body diagram for the boxes.

Each box has mass, so each box has a gravitational force. Since we don't know that the boxes have the same mass, we use m_1 and m_2 for the masses and $m_1 g$ and $m_2 g$ for the gravitational forces. Each box is in contact with the floor, so each could have a normal force acting on it. They might not have the same magnitude, so we use N_1 and N_2 for the normal forces.

Joe pushes the left box to the right, and if there are no other horizontal forces it will accelerate to the right. What keeps the two boxes from occupying the same space? The left box pushes the right box to the right, so that the boxes accelerate together. We label this force F_{BX}.

Newton says that if the box 1 pushes box 2 to the right then box 2 pushes box 1 to the left. This force is equal as well as opposite, so it has the same magnitude as F_{BX}. We draw a force on the left box going to the left, labeled F_{BX}.

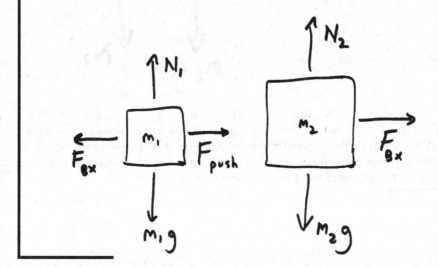

EXAMPLE

Box 1 sits on box 2, which sits on the floor. Draw a free body diagram for the boxes.

Each box has mass, so each box has a gravitational force. Since the boxes have different masses, we need different gravitational forces $m_1 g$ and $m_2 g$.

Box 1 is in contact with box 2, so there is a normal force N_1 pushing on box 1. This force is perpendicularly out of the top of box 2, or upward. As usual, we don't know the magnitude of the normal force.

Box 2 is in contact with the floor, so there is a normal force pushing box 2 upward. This force is not necessarily the same as N_1 pushing on box 1, so we call it something different, N_2. Box 2 is also in contact with box 1. Since box 2 pushes up on box 1 with force N_1, box 1 pushes down on box 2 with force N_1.

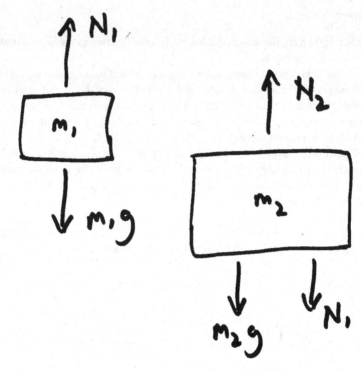

It is not true that the weight of box 1 acts on box 2. There is a normal force from box 1 pushing down on box 2, and sometimes this force is equal to the weight of box 1, but it isn't always equal to the weight of box 1. If the boxes and floor are accelerating up or down, for example, then the normal force N_1 will not be equal to the weight. The weight of box 1 acts only on box 1.

LESSON

Newton's third-law equal and opposite forces never act on the same object.

3.2 SOLVING $F = ma$

Newton's second law says that if we add all of the forces on something we will get the mass times the acceleration. Now that we have a complete list of the forces (the free body diagram), we want to add them. Keep in mind that some of the forces may not be known, but we can add them in as symbols and maybe solve for them later.

Whenever we are dealing with force, we start by drawing a free body diagram. The next step is to choose a coordinate system. If all of the forces are colinear (parallel or antiparallel), then this is the same as choosing the positive direction in a constant acceleration problem. Often the forces won't be colinear, and we'll need to pick a pair of axes (or sometimes a trio). If so then it is vital that the axes be perpendicular to each other, so that all of our vector techniques will work. Any set of perpendicular axes will work, but some will be easier than others. The math will be much easier to solve if the acceleration is parallel to one of the axes.

A good set of axes needs these features:

- One axis suffices if all forces are parallel, two if all forces are in the same plane, otherwise three axes.
- Each axis must be perpendicular to every other axis.
- Each axis needs a unique name (such as x or y).
- Each axis needs to have the positive direction clearly marked.
- If possible, try to line up one axis parallel to the acceleration.

After choosing axes we take each vector and divide it into components. We then add all of the components that are parallel to one of the axes. We set this sum equal to the mass times the acceleration along that axis. Hopefully we'll have an equation we can solve, but if not we can try the other axis.

EXAMPLE

A boy pulls a 12 kg sled along an icy, frictionless ground with a rope. The rope is 29° above horizontal and has 18 N of tension. What is the acceleration of the sled?

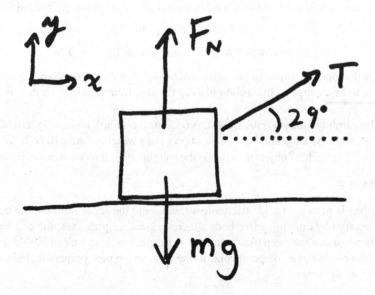

We start by drawing a free body diagram. Next we pick axes: the acceleration of the box is to the right, so we call this the x-axis. We need a second axis, perpendicular to the first, so we pick upward as the y-axis.

Since we want the acceleration and it is in the x direction, we add the forces in the x direction.

$$F = ma$$

$$F_x = ma_x$$

$$+T \cos 29° = ma_x$$

Only part of the tension is in the x direction, so we have to find the x component. The part we want is adjacent to the angle, so it's the cosine of $29°$. The plus sign in front of the T indicates that the x component of T is in the $+x$ direction rather than the $-x$ direction.

We can solve this equation to get the acceleration.

$$+(18 \text{ N}) \cos 29° = (12 \text{ kg})a_x$$

$$a_x = \frac{(18 \text{ N}) \cos 29°}{(12 \text{ kg})} = +1.3 \text{ m/s}^2$$

Note that the N in (18 N) is the unit newtons, as opposed to the variable N which is the magnitude of the normal force. Now we add the forces in the y direction (mostly for practice).

$$F_y = ma_y$$

$$+T \sin 29° + N - mg = ma_y = 0$$

Remember that T, N, and mg are magnitudes, so the signs in front of them indicate the direction of the y component of the force.

Since we know the tension, we can solve that equation for the normal force.

$$+N = mg - T \sin 29°$$

$$+N = (12 \text{ kg})(9.8 \text{ m/s}^2) - (18 \text{ N}) \sin 29° = 109 \text{ N}$$

Because the rope pulls upward (as well as sideways) on the sled, the normal force does not have to be as large as the weight to keep the sled from accelerating downward.

Be careful of mass versus weight. A 6 kg mass on Earth has a weight of 60 N. If it is taken to the Moon it still has a mass of 6 kg but a weight of only 10 N. If an object is called a "30 newton" object then that is the weight rather than the mass of the object.

EXAMPLE

A 6 kg box is placed on a $32°$ frictionless ramp. Find the acceleration of the box.

We start by drawing a free body diagram. Next we pick axes: the acceleration of the box is down the ramp (down and left in the drawing), so we call this the x-axis. We need a second axis, perpendicular to the first, so we pick perpendicularly out of the ramp as the y-axis.

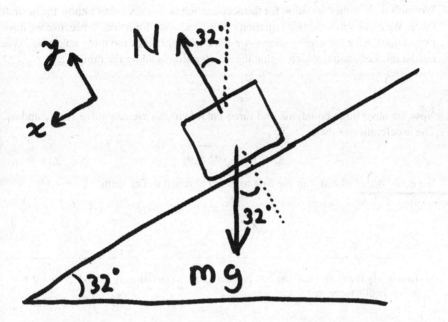

Since we want the acceleration and it is in the x direction, we add the forces in the x direction.

$$F = ma$$
$$F_x = ma_x$$
$$+mg \sin 32° = ma_x$$
$$a_x = (9.8 \text{ m/s}^2) \sin 32° = 5.2 \text{ m/s}^2$$

The mass cancels, so the acceleration doesn't depend on the mass. How can this be? If the box has twice the mass, it has twice the gravitational force, but it has twice the mass to move, so the acceleration is the same. That is why everything falls with the same acceleration.

Now we add the forces in the y direction (again for practice).

$$F_y = ma_y$$
$$+N - mg \cos 32° = ma_y = 0$$
$$N = mg \cos 32° = (6 \text{ kg})(9.8 \text{ m/s}^2) \cos 32° = 50 \text{ N}$$

What if we hadn't used these axes, but instead used the more traditional x to the right and y upward?

$$F_x = ma_x$$
$$-N \sin 32° = ma_x$$
$$F_y = ma_y$$
$$+N \cos 32° - mg = ma_y$$

We can't solve either equation for the acceleration because we don't know the normal force. We can't solve the two equations simultaneously for a and N because we have two variables for the acceleration—we have two equations but three unknowns. We need a third equation, which is that the acceleration is along the ramp.

$$\frac{a_y}{a_x} = \tan 32°$$

Now we have three equations and three unknowns, so we can solve for a_x and a_y. The acceleration is then

$$a = \sqrt{a_x^2 + a_y^2}$$

We could do all of this, but the first axes led to much easier math.

LESSON

To simplify the math, pick one axis parallel to the acceleration. A second axis must be perpendicular to the first.

EXAMPLE

Two boxes are attached by a massless rope that hangs over a massless, frictionless pulley. The masses of the boxes are 10 kg and 20 kg. What is the acceleration of each box? (This is known as an Atwood's machine.)

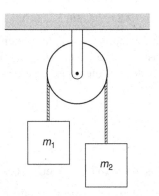

Each box has mass, so gravity pulls on each box. The masses are not the same, so we label them m_1 and m_2, where $m_1 = 10$ kg and $m_2 = 20$ kg. Each box also has a rope pulling it upward, so there is a tension force on each box. These tension forces are the same as long as the rope and pulley are both massless and frictionless, so we label them each T. (For the tensions to be the same, we need only the rope to be massless and any one of the other conditions to be true—we'll tackle this again later.)

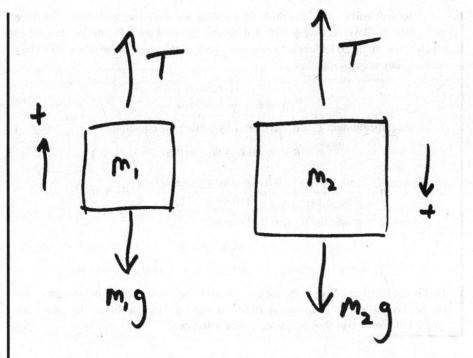

Since the right-hand mass is heavier, we expect that m_2 accelerates down while m_1 accelerates up. Therefore we choose these as our positive direction: positive is upward on the left side and downward on the right side. Can we do this, having two different coordinate systems in the same problem? Yes, each object can have its own coordinate system, but we must be careful to keep them straight.

Before we begin our calculation, consider the possible values for T. If T is equal to the weight of the 10 kg box, then the total force on that box is zero, so it doesn't accelerate. But the total force on the 20 kg box will be downward, so that box does accelerate. Could these both be true? No, so the magnitude of the tension force is between the weights of the two boxes (not necessarily halfway between).

Now we add the forces on the 10 kg box.

$$F = ma$$
$$+T - m_1g = m_1a_1$$

Can we solve this to find the acceleration? No, because we don't know the tension. So let's try the 20 kg box.

$$F = ma$$
$$+m_2g - T = m_2a_2$$

Remember that T and mg are magnitudes, so the signs in front of them indicate the direction of the forces. For the right-hand box, downward is the positive direction. The gravitational force is downward so it is positive, and the tension is upward so it is negative. T and mg must be positive values since they are magnitudes.

We still can't solve the problem because we have two equations and three unknowns. The last piece we need is that the accelerations are the same: the downward acceleration of the 20 kg box is the same as the upward acceleration of the 10 kg box because they are tied together.

$$\begin{cases} +T - m_1g = m_1a \\ +m_2g - T = m_2a \end{cases}$$

We can eliminate T and solve for a by adding the equations.

$$m_2g - m_1g = m_1a + m_2a$$

$$a = \frac{(m_2 - m_1)g}{(m_1 + m_2)} = \frac{(20 \text{ kg} - 10 \text{ kg})(9.8 \text{ m/s}^2)}{(10 \text{ kg} + 20 \text{ kg})} = 3.3 \text{ m/s}^2$$

We can now find the tension in the rope.

$$+T - m_1g = m_1a$$

$$T = m_1a + m_1g = (10 \text{ kg})(3.3 \text{ m/s}^2 + 9.8 \text{ m/s}^2) = 131 \text{ N}$$

This is not halfway between the weights of the boxes, but closer to the weight of the smaller box. The total force on the right box must be twice as big as the total force on the left box so that they accelerate at the same rate.

EXAMPLE

Two boxes of masses 20 kg and 40 kg rest on a frictionless floor. Joe pushes the left box to the right with a force of 17 N. Find the acceleration of each box.

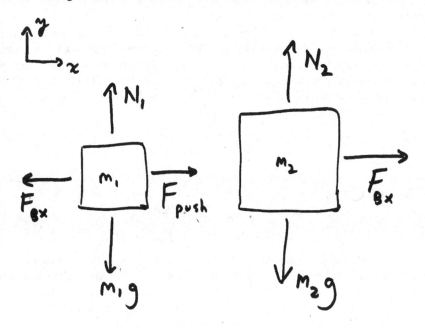

We start by drawing a free body diagram. Next we pick axes: the acceleration of the boxes will be to the right, so call this the x-axis. We need a second axis, perpendicular to the first, so pick upward as the y-axis.

Since we want the acceleration and it is in the x direction, we add the forces in the x direction on the left box (m_1).

$$F = ma$$
$$F_x = ma_x$$
$$+17\,\text{N} - F_{BX} = m_1 a_{1,x}$$

We can't solve this because we don't know F_{BX}.

We add the forces in the x direction on the right box (m_2).

$$F_x = ma_x$$
$$+F_{BX} = m_2 a_{2,x}$$

We can't solve this equation either.

But the accelerations of the two boxes will be the same. If they weren't, then either the left box accelerates into the right box or the right box accelerates away from the left box. So

$$\begin{cases} +17\,\text{N} - F_{BX} = m_1 a \\ \quad\quad +F_{BX} = m_2 a \end{cases}$$

We add the equations.

$$17\,\text{N} = (m_1 + m_2)\,a$$

This is the same thing we would have gotten if we had treated the two boxes as one object. As long as they always move together, we can treat them as a single big box.

$$a = \frac{17\,\text{N}}{(m_1 + m_2)} = \frac{17\,\text{N}}{20\,\text{kg} + 40\,\text{kg}} = 0.28\,\text{m/s}^2$$

LESSON

When two objects act on one another, Newton's third law lets us set the two forces equal to each other.

EXAMPLE

A 6 kg box sits on a 40° frictionless ramp. It is attached by a massless, frictionless rope to a 4 kg box. What is the acceleration of each box?

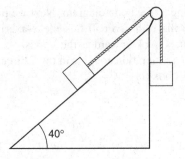

Each box has mass, so gravity pulls on each box. The masses are not the same, so we label them m_1 and m_2, where $m_1 = 6\,\text{kg}$ and $m_2 = 4\,\text{kg}$. There is a rope pulling on each box, so there is a tension force on each box. These tension forces are the same as long as the rope is both massless and frictionless, so we label them both T. The left box is in contact with a surface, the ramp, so we include the unknown normal force N.

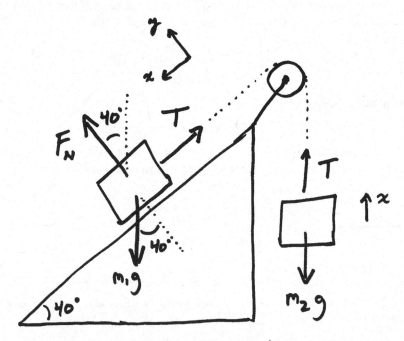

The acceleration of the left box will be along the ramp while the acceleration of the right box will be vertical. Which way will they be? If the right box is much heavier than the left then it will accelerate downward, but if it is much lighter then it will accelerate upward. Since our two masses are comparable (about the same), it is not immediately obvious which way they will go.

Which direction should we pick as our x-axis? If we pick down the ramp, but the acceleration of m_1 is up the ramp, then we will get a negative value. Getting a negative value was a common occurrence in the last chapter—there is no problem as long as we use the axes consistently. With this in mind, we'll use down the ramp as the positive x direction on m_1 and up as the positive x direction on m_2. That way they both accelerate in the same direction, either $+x$ or $-x$.

Now we add the forces on m_1 in the x direction.

$$F_x = ma_x$$

$$+m_1 g \sin 40° - T = m_1 a_x$$

We can't solve this because we don't know the tension, so we do the same for m_2. For m_2 the x direction is vertically up.

$$F_x = ma_x$$

$$+T - m_2 g = m_2 a_x$$

We can eliminate T and solve for a_x by adding the equations.

$$m_1 g \sin 40° - m_2 g = m_1 a_x + m_2 a_x$$

$$a_x = \frac{(m_1 \sin 40° - m_2)g}{(m_1 + m_2)} = \frac{[(6 \text{ kg}) \sin 40° - (4 \text{ kg})](9.8 \text{ m/s}^2)}{(6 \text{ kg} + 4 \text{ kg})} = -0.14 \text{ m/s}^2$$

The negative sign in our result means that the acceleration is up the ramp (m_1) and vertically down (m_2). Does this mean that our choice of axes was wrong? No, either choice would have been equally good. It only means that the acceleration is in the opposite direction as what we chose as positive.

CHAPTER SUMMARY

Whenever dealing with forces:

- Draw a free body diagram.
- Pick one axis parallel to the acceleration.
- Add the forces in each direction and set it equal to ma.
- Solve from there.

Remember:

- The normal force is perpendicular to the surface.
- We don't know the magnitude of the normal force—it isn't always equal to the weight.
- Newton's third-law "equal and opposite" forces never act on the same object.

CHAPTER 4

FORCES, THE SEQUEL

Forces are too big of a topic to cover in a single chapter. The goal of the last chapter was to develop the correct technique using Newton's laws. In this chapter we'll build on this by adding a new force and a nonforce.

Friction can be a little tricky. In particular, the direction of the friction force is not always obvious. When drawing a free body diagram, do friction last. Each time you come to a normal force there could be an associated friction force. Make a note of that (mentally or on paper), do all of the other forces, then come back to do the friction force.

The other topic in this chapter is circular motion and centripetal acceleration. If you are in a car and the car turns to the left, you might feel a "force" to the right. Some people call this the centripetal force. Such a force does not really exist, as we'll see.

Consider what happens from the view of a nearby pedestrian. If you are driving forward and stop, you might think that you are thrown forward. The pedestrian knows that you are continuing in a straight line while the car slows down around you. When you turn the car to the left, you might think that you are thrown to the right. The pedestrian knows that you are continuing in a straight line while the car turns underneath you.

4.1 FRICTION FORCES

Whenever an object is in contact with a surface, there can be a normal force. Whenever there is a normal force, there can be a friction force. The friction force is always parallel to the surface.

Friction does not oppose motion—friction opposes sliding of surfaces against one another. Imagine that you are loading bags of sand into the back of a pickup truck. You toss the bag forward into the truck bed, and when it hits the truck it is moving forward. It slides against the truck bed, but friction opposes this sliding, slowing and stopping the bag. After the truck is loaded you drive it forward. If there were no friction between the truck bed and the bag of sand then the bag would stay in place while the truck moves out from underneath it. Friction pushes the bag forward with the truck—friction causes the motion that prevents sliding.

Friction occurs in two types: kinetic and static. **Kinetic friction** occurs when two surfaces are sliding against each other. The friction force that slows the bag as it is loaded is kinetic friction. **Static friction** occurs when surfaces are not yet sliding against each other. The friction force that accelerates the bag of sand with the truck is static friction. Static friction tries to prevent sliding while kinetic friction tries to stop sliding that is already happening. The two types of friction have different names because they act differently.

The magnitude of kinetic friction is

$$\mathcal{F}_k = \mu_k N$$

where N is the magnitude of the normal force between the two surfaces. The symbol μ_k is the coefficient of kinetic friction, a constant that depends on the two surfaces. The direction of the kinetic friction force is always to slow the sliding between the surfaces.

The direction of a static friction force is always to prevent sliding between the surfaces. The magnitude of static friction is as big as it needs to be to prevent sliding, up to some limit. The magnitude of the static friction force is

$$\mathcal{F}_s \leq \mu_s N$$

where N is the magnitude of the normal force between the two surfaces and μ_s is the coefficient of static friction.

Consider again the bag of sand in the truck. After the bag has been loaded but before the truck starts moving, the bag sits in the back of the truck without sliding. Since it is not sliding (yet), the friction force is static friction. How much static friction force is needed to keep it from sliding? There are no horizontal forces on the bag other than friction, so no friction force is needed. If there were a friction force it would cause the bag to move and to start sliding on the truck bed. So the friction force is zero until the truck tries to move.

We covered forces and free body diagrams a couple chapters ago, so why did we save friction until now? When surfaces are already slipping and the friction is kinetic, it is not hard to determine the direction of the friction force. When the surfaces are not yet slipping and the friction is static, it can be tricky to determine the direction of the static friction force. It is best to include all other forces in the free body diagram before putting in the friction.

EXAMPLE

You toss an 8 kg bag of sand into the back of a pickup truck. The sand hits the truck bed moving horizontally with a speed of 1.2 m/s. The coefficient of kinetic friction between the bag and the truck is 0.46. How far does the bag of sand slide before coming to a stop?

We know the initial and final velocities. If we knew the (constant) acceleration then we could use our "three of five" technique to find the displacement. To find the acceleration we add the forces and use Newton's second law.

The bag of sand has mass so there is a gravitational force. It is in contact with the truck bed so there could be a normal force. It is sliding toward the front of the truck so there is a kinetic friction force pushing the bag toward the back of the truck, against the sliding.

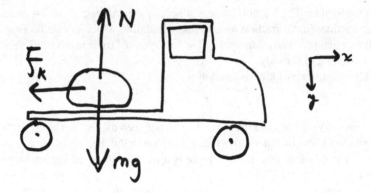

The acceleration is left along the truck bed; we'll use to the right as the x-axis and down as the y direction. We want to find a_x so we add the forces in the x direction.

$$F_x = ma_x$$
$$-\mathcal{F}_k = ma_x$$

Because the friction is kinetic friction, the magnitude of the friction force is equal to the coefficient of friction times the normal force.

$$\mathcal{F}_k = \mu_k N \qquad \rightarrow \qquad -\mu_k N = ma_x$$

We need to find the normal force N, so we add the forces in the y direction.

$$F_y = ma_y$$
$$+mg - N = ma_y = m(0) = 0$$
$$N = mg$$

Because the x-axis is parallel to the acceleration, the acceleration in the y direction is zero. There are only two forces in the y direction and no acceleration, so the forces are equal and cancel each other out.

$$-\mu_k(mg) = ma_x$$
$$a_x = -\mu_k g = -(0.46)(9.8 \text{ m/s}^2) = -4.5 \text{ m/s}^2$$

Why does the mass cancel? If the mass of the bag were doubled, its weight would double and the normal force would double, so the friction force would double, but that force would act on twice as much mass so the acceleration would be the same. Now we can do the constant acceleration problem:

$$x = ?$$
$$v_0 = +1.2 \text{ m/s}$$
$$v = 0$$
$$a = -4.5 \text{ m/s}^2$$
$$t =$$

We don't care about the time so we choose $v^2 - v_0^2 = 2a \, \Delta x$.

$$v^2 - v_0^2 = 2a \, \Delta x$$

$$\Delta x = \frac{v^2 - v_0^2}{2a}$$

$$\Delta x = \frac{(0)^2 - (1.2 \text{ m/s})^2}{2(-4.5 \text{ m/s}^2)}$$

$$\Delta x = 0.16 \text{ m}$$

EXAMPLE

A 8 kg bag of sand sits in the back of a pickup truck. The coefficient of static friction between the bag and the truck is 0.56. What is the maximum acceleration the truck can have so that the bag does not slip on the truck bed?

We add the forces and use Newton's second law to find the acceleration. The bag of sand has mass so there is a gravitational force. It is in contact with the truck bed so there could be a normal force. Because there is a normal force, there could be a friction force. The bag is not yet sliding on the truck bed so any friction force is static friction.

The truck will accelerate to the right, so the static friction force will push the bag to the right to keep it with the truck. We draw the friction force to the right and choose to the right as the x direction.

$$F_x = ma_x$$

$$+\mathcal{F}_{\text{s}} = ma_x$$

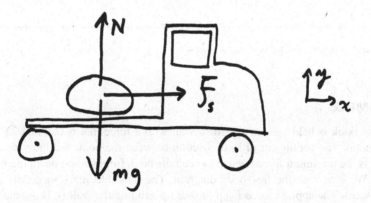

The maximum acceleration occurs when the friction force is the biggest it can get. The biggest the friction force can get is

$$ma_x = \mathcal{F}_{\text{s}} \leq \mu_{\text{s}} N$$

We need to find the normal force N, so we add the forces in the y (up) direction.

$$F_y = ma_y$$
$$+N - mg = ma_y = m(0) = 0$$
$$N = mg$$
$$ma_x \leq \mu_s mg$$
$$a_x \leq \mu_s g = (0.56)(9.8 \text{ m/s}^2) = 5.5 \text{ m/s}^2$$

The truck itself is accelerated by static friction, this time between the tires and the road. The tires are rotating, but as long as they don't slip on the road the friction is static friction.

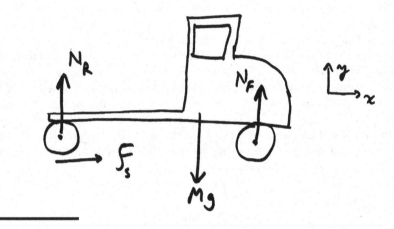

EXAMPLE

A 5 kg book is held against a vertical wall with a force that is directed 25° above horizontal. The coefficient of static friction between the book and the wall is 0.73. What is the minimum force needed to keep the book from sliding down the wall?

We start with the free body diagram. The book has mass so gravity pulls it downward. The applied force F_{hand} pushes up and into the wall (it is the magnitude of this force that we're trying to find). Because the book is in contact with a surface (the wall) there is a normal force perpendicularly out of the wall, or horizontal. We do not know how big this force is, so we call it F_N.

Whenever there is a normal force there could be a friction force. The friction force is parallel to the surface, so it could push the book up or down. The friction force will be whatever it needs to be, up to the limit $\mu_s F_N$, so that the book does not slide. If we press hard and mostly upward on the book it would slide up the wall, so the friction force would be downward. We want to keep the book up by applying only a little force, so we want friction to help hold the book up. Friction pushes the book up the wall, helping us to hold it there.

We choose coordinate axes: there is no acceleration so we can pick anything we want. We'll use x right and y up.

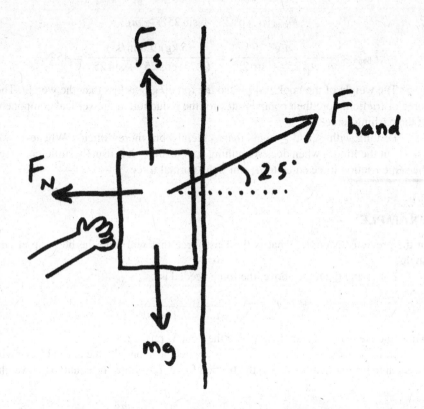

Now we add the forces along each axis, setting the sum equal to the mass of the book times its acceleration along that axis.

$$F_x = ma_x$$

$$+F_{\text{hand}} \cos 25° - F_N = ma_x = 0$$

$$F_y = ma_y$$

$$+F_{\text{hand}} \sin 25° - mg + \mathcal{F}_s = ma_y = 0$$

We have two equations and three unknowns (F_{hand}, F_N, and $\mathcal{F}_s$). We also know that the limit to the static friction force is $\mu_s F_N$, so

$$-F_{\text{hand}} \sin 25° + mg = \mathcal{F}_s \leq \mu_s F_N$$

We find the normal force from the x equation.

$$F_N = F_{hand} \cos 25°$$

$$-F_{hand} \sin 25° + mg \leq \mu_s F_{hand} \cos 25°$$

$$mg \leq \mu_s F_{hand} \cos 25° + F_{hand} \sin 25°$$

$$mg \leq F_{hand}(\mu_s \cos 25° + \sin 25°)$$

$$F_{hand}(\mu_s \cos 25° + \sin 25°) \geq mg$$

$$F_{hand} \geq \frac{mg}{\mu_s \cos 25° + \sin 25°} = \frac{(5 \text{ kg})(9.8 \text{ m/s}^2)}{(0.73) \cos 25° + \sin 25°} = 45 \text{ N}$$

The weight of the book is 49 N, so the force F_{hand} is less than the weight. The force of friction more than compensates for the reduction in the vertical component of the pushing force.

Dealing with $\leq$ and $\geq$ signs makes the algebra more difficult. Whenever we ask about the limit—when does something start to slip—it is usually sufficient to set the static friction force equal to μ_s times the normal force.

EXAMPLE*

In the previous example, what is the least force that will hold the book up at any angle?

Following the logic above, the force needed is

$$F_{min}(\theta) = \frac{mg}{\mu_s \cos \theta + \sin \theta}$$

where the angle θ is the angle at which the book is pushed.

If the angle is zero, then the force is applied horizontally, the normal force will be equal to the applied force, and the friction force $\mu_s F_N$ must be equal to the weight mg,

$$mg = \mu_s F_N = \mu_s F \quad \rightarrow \quad F = \frac{mg}{\mu_s}$$

which matches our equation. If the angle θ is 90° then the force is applied vertically and we expect that $F = mg$, which also matches our equation. Since the coefficient of friction is usually less than one, shouldn't the minimum force be when θ is 90°? Not necessarily. In our example just above the minimum force was less than the weight. Because $\cos \theta$ does not change much as the angle increases from zero, the increase in the horizontal component of the force is greater than the drop in the vertical component of the force as θ increases from zero. For a while the increased friction force more than makes up for the decreased vertical applied force.

*Example uses calculus.

The question is when does the increase in the friction force equal the decrease in the vertical component of the applied force? This happens when the change in the minimum force needed is zero as the angle increases, or when

$$\frac{d}{d\theta}F_{\text{HAND}} = 0$$

$$\frac{d}{d\theta}\left(\frac{mg}{\mu_s \cos\theta + \sin\theta}\right) = 0$$

$$-\frac{(mg)}{(\mu_s \cos\theta + \sin\theta)^2}\frac{d}{d\theta}(\mu_s \cos\theta + \sin\theta) = 0$$

$$-\mu_s \sin\theta + \cos\theta = 0$$

$$\frac{\sin\theta}{\cos\theta} = \frac{1}{\mu_s}$$

$$\tan\theta = \frac{1}{\mu_s}$$

Putting this angle back into the equation for the force (and using some trig identities) gives a minimum force of

$$F_{\min} = \frac{mg}{\sqrt{\mu_s^2 + 1}}$$

For our case of $\mu_s = 0.73$, the minimum force is 81% of the weight when the angle θ is 54° above horizontal.

EXAMPLE

A block is sent sliding up a 50° ramp at an initial speed of 8 m/s. The coefficient of kinetic friction between the ramp and the block is 0.46. How far up the ramp does the block go before stopping?

The forces are constant, so the acceleration should be constant. We can apply the constant acceleration technique if we know three of the five symbols. We know the initial and final velocities ($v_f = 0$), and we want to find the displacement, but we don't know either the time or the acceleration. If we could find the acceleration by adding the forces then we could do the constant acceleration problem.

To add the forces, we start by drawing a free body diagram. The block has mass, although we aren't told what it is. We'll use m for the mass; maybe we won't need it. The weight is $W = mg$ downward. The block is in contact with a surface so there is a normal force N perpendicularly out of the surface. Because there is a normal force there could be a friction force. Since the block is sliding up the ramp there is a kinetic friction force F_{fr} down the ramp.

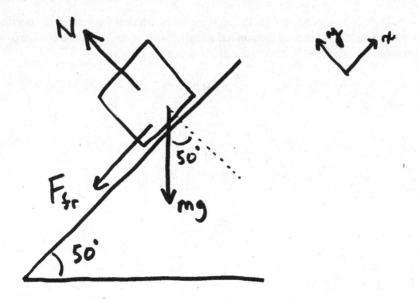

What about the force that pushes the block up the ramp? That force started the block, but it is no longer acting on the block. We want only the forces on the block as it slides up the ramp, after it has been given its initial 8 m/s speed.

To add the forces we need to choose a pair of coordinate axes. The block is moving up the ramp and slowing, so the acceleration is down the ramp. We want one axis to be parallel to the acceleration, so x is either up or down the ramp. If we pick down the ramp as $+x$ then the acceleration will be positive, while if we pick up the ramp as $+x$ then the initial velocity and displacement will be positive. Since we're trying to find the displacement, we'll use up the ramp as the $+x$ axis.

We add the forces in the x direction.

$$F_x = ma_x$$
$$-mg\sin 50° - F_{fr} = ma_x$$
$$-mg\sin 50° - \mu_k N = ma_x$$

We want to solve this for a_x, but we don't know m or N. To find N we add the forces in the y direction.

$$F_y = ma_y$$
$$+N - mg\cos 50° = ma_y = 0$$

We chose the x-axis to be parallel to the acceleration so the y acceleration is zero.

$$N = mg\cos 50°$$
$$-mg\sin 50° - \mu_k(mg\cos 50°) = ma_x$$

The mass cancels.

$$a_x = -g \sin 50° - \mu_k g \cos 50°$$
$$a_x = -g (\sin 50° + \mu_k \cos 50°)$$
$$a_x = -10.4 \text{ m/s}^2$$

Now we can do the constant acceleration part.

$$\Delta x = ?$$
$$v_0 = +8\text{m/s}$$
$$v = 0$$
$$a = -10.4\text{m/s}^2$$
$$t =$$

We don't care about the time it takes for the block to stop, so

$$v^2 - v_0^2 = 2a \, \Delta x$$
$$\Delta x = \frac{v^2 - v_0^2}{2a}$$
$$\Delta x = \frac{(0)^2 - (+8 \text{ m/s})^2}{2(-10.4 \text{ m/s}^2)}$$
$$\Delta x = 3.08 \text{ m}$$

If the block starts back down the ramp, the friction force will point up the ramp. The acceleration will not be the same as it was as the block went up the ramp. We would need to draw a new free body diagram and add the forces again.

LESSON

Friction forces are always parallel to the surfaces.

EXAMPLE

A small box ($m = 2.0$ kg) is in contact with a larger box ($m = 18$ kg) as shown in the picture. A force $\mathcal{F}$ pushes on the larger box. Because of friction, if the force $\mathcal{F}$ is large enough the small box will not fall. How large does $\mathcal{F}$ need to be? The coefficients of friction at all surfaces are $\mu_s = 0.62$ and $\mu_k = 0.42$.

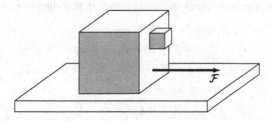

As always, we start with the free body diagrams. We need a separate diagram for each box (we'll see why in a moment). For the small box, we have the weight mg downward. It is in contact with the large box, so there could be a normal force N to the right. Because there is a normal force there could be a friction force. If there were no friction then the small box would slide down the large box. To prevent this there is a static friction force f pushing the small box upward.

The large box has mass so gravity acts on it. The masses of the boxes are not the same, so we call the gravity force on the large box Mg. An invisible hand pushes it to the right with a force $\mathcal{F}$. It is in contact with the floor so there could be a normal force. Since we've already used N for the force between the boxes, and this force is not necessarily the same as that force, we'll call the normal force of the floor on the large box η. Because there is a normal force there could be a friction force F_k, pushing the large box to the left as it slides to the right.

The large box is also in contact with the small box. If the large box pushes the small box to the right with a normal force N, then the small box pushes the large box to the left with a normal force N. If the large box pushes the small box upward with a friction force f, then the small box pushes the large box downward with a friction force f. This is Newton's third law.

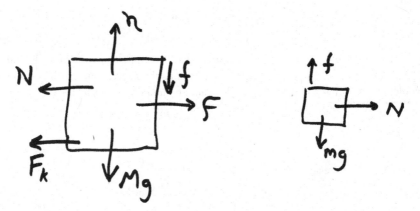

If we had used a single free body diagram for the two combined objects, then the forces that they apply to each other (N and f) cancel. They wouldn't appear in our equations, but we need them because it is the friction force that holds up the small box.

Since the acceleration is horizontal we pick horizontal and vertical as our axes. We use Newton's second law, adding the forces on each box in each direction and setting it equal to the mass times the acceleration in that direction.

$$F = ma$$
$$+\mathcal{F} - N - F_k = Ma_x$$
$$+\eta - Mg - f = Ma_y = 0$$
$$+N = ma_x$$
$$+f - mg = ma_y = 0$$

We use the same acceleration a_x for both boxes because they move together. The vertical acceleration a_y is zero because we chose the x axis to be parallel to the acceleration.

Because we want the limiting value of $\mathcal{F}$, we want the limiting value of the static friction force, so $f = \mu_s N$, where N is the normal force between the two boxes. The kinetic friction between the large box and the floor is $\mu_k \eta$, where η is the normal force between the large box and the floor.

$$\begin{cases} +\mathcal{F} - N - \mu_k \eta = Ma_x \\ +\eta - Mg - \mu_s N = 0 \\ \qquad +N = ma_x \\ +\mu_s N - mg = 0 \end{cases}$$

This may look complicated, but it's not so bad. From the fourth equation we learn that the static friction f is equal to the weight of the small box, so we can find N. From the third equation we find the acceleration. From the second equation we find the normal force between the large box and the floor. We put everything into the first equation and solve for $\mathcal{F}$.

$$N = \frac{mg}{\mu_s}$$

$$a_x = \frac{N}{m} = \frac{mg}{\mu_s}\frac{1}{m} = \frac{g}{\mu_s}$$

$$\eta = Mg + \mu_s N = Mg + mg = (M + m)g$$

$$\mathcal{F} = N + \mu_k \eta + Ma_x$$

$$\mathcal{F} = \frac{mg}{\mu_s} + \mu_k(M + m)g + M\frac{g}{\mu_s}$$

$$\mathcal{F} = (M + m)\left(\frac{1}{\mu_s} + \mu_k\right)g$$

$$\mathcal{F} = (18 \text{ kg} + 2.0 \text{ kg})\left(\frac{1}{0.62} + 0.42\right)(9.8 \text{ m/s}^2)$$

$$\mathcal{F} = 398 \text{ N}$$

EXAMPLE

A small box ($m = 2.0$ kg) rests on a larger box ($m = 18$ kg) as shown. A force $\mathcal{F}$ pushes on the smaller box. What is the maximum force $\mathcal{F}$ such that the small block does not slip on the large block? The coefficients of friction at all surfaces are $\mu_s = 0.62$ and $\mu_k = 0.42$.

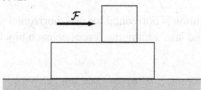

We begin by drawing the free body diagrams. We use a separate diagram for each block. For the small block, we have the weight mg downward and the applied force $\mathcal{F}$ to the right. It is in contact with the large block, so there could be a normal force N upward. Because there is a normal force there could be a friction force. If there were no friction then the large block could not move horizontally—there are no horizontal forces on it except friction. The small block would slide off of the large one. To prevent this there is a static friction force f pushing the large block to the right and equal and opposite force f pushing the small block to the left.

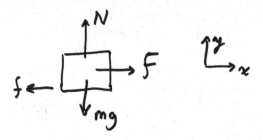

The large block has mass so gravity pulls it downward with a force Mg. It is in contact with the floor so there could be a normal force. Since we've already used N, and since this force is not necessarily the same as that force, we'll call the normal force of the floor on the large block η. Since the large block pushes up on the small one with a force N, the small one pushes down on the large one with a force N. As we saw in the last paragraph, there is a static friction force f pushing the large block to the right, but there is also a kinetic friction force F_k pushing the large block to the left.

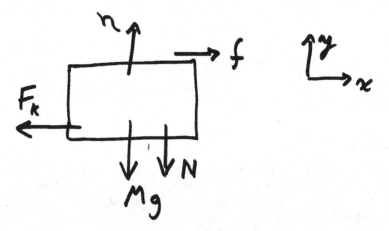

Since the acceleration is horizontal we pick horizontal and vertical as our axes. We use Newton's second law, adding the forces on each block in each direction.

$$F = ma$$
$$+\mathcal{F} - f = ma_x$$

$$-mg + N = ma_y = 0$$
$$-F_k + f = Ma_x$$
$$+\eta - N - Mg = Ma_y = 0$$

We use the same horizontal acceleration a_x for both blocks because they move together. The vertical acceleration a_y is zero because we picked the x-axis to be parallel to the acceleration.

Because we want the limiting value of $\mathcal{F}$, we want the limiting value of the static friction force, so $f = \mu_s N$, where N is the normal force between the two blocks. The kinetic friction between the large block and the floor is $\mu_k \eta$, where η is the normal force between the large block and the floor.

$$\begin{cases} +\mathcal{F} - \mu_s N = ma_x \\ -mg + N = ma_y = 0 \\ -\mu_k \eta + \mu_s N = Ma_x \\ +\eta - N - Mg = Ma_y = 0 \end{cases}$$

From the second equation we learn that the normal force N between the blocks is equal to the weight of the small block mg (it's our special case: two vertical forces and no vertical acceleration). From the fourth equation we find the normal force η between the large block and the floor. From the third equation we find the acceleration. We put everything into the first equation and solve for $\mathcal{F}$.

$$N = mg$$
$$\eta = N + Mg = (m + M)g$$
$$-\mu_k(m + M)g + \mu_s mg = Ma_x$$
$$a_x = \frac{\mu_s mg}{M} - \frac{\mu_k(m + M)g}{M}$$
$$\mathcal{F} = \mu_s mg + ma_x$$
$$\mathcal{F} = \mu_s mg + \frac{\mu_s m^2 g}{M} - \frac{\mu_k(m + M)mg}{M}$$
$$\mathcal{F} = \left(\frac{\mu_s mM + \mu_s m^2 - \mu_k(m + M)m}{M} \right) g$$
$$\mathcal{F} = \left(\frac{(0.62)(2 \text{ kg})(18 \text{ kg}) + (0.62)(2 \text{ kg})^2 - (0.42)(20 \text{ kg})(2 \text{ kg})}{(18 \text{ kg})} \right) g$$
$$\mathcal{F} = 4.4 \text{ N}$$

LESSON

If we care about the forces that two objects put on one another, then we need separate free body diagrams.

EXAMPLE

A refrigerator with a weight of 500 N sits on the floor. The coefficient of kinetic friction between the refrigerator and the floor is 0.30 and the coefficient of static friction is 0.40. A force of 170 N is applied horizontally to the refrigerator. What is the magnitude of the friction force the floor applies to the refrigerator?

The refrigerator has mass so there is weight downward. It is in contact with the floor so there is a normal force upward. Because there is a normal force there could be a friction force f, which could be left or right. Someone pushes the refrigerator sideways with a force P, so we'll draw that to the right. Friction must push to the left to prevent the refrigerator from sliding.

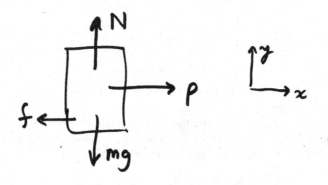

We pick the x-axis to the right and add the forces.

$$F_x = ma_x$$
$$P - f = ma_x$$

If P is greater than the friction force then the refrigerator will accelerate to the right, sliding across the floor. In this case the friction will be kinetic and equal to $\mu_k N$. If the friction force is greater than P then the refrigerator will accelerate to the left—someone pushes to the right and the refrigerator comes back toward him because of the overpowering friction! That doesn't happen. Friction will be no stronger than his push on the refrigerator. If static friction is strong enough to keep the refrigerator from sliding, then it will be just strong enough to do so and no stronger.

The maximum static friction force is

$$f_{max} = \mu_s N$$

We add the vertical forces.

$$+N - mg = ma_y = 0$$
$$N = mg$$
$$f_{max} = (0.40)(500 \text{ N}) = 200 \text{ N}$$

Friction can push back with up to 200 N, more than enough to stop the refrigerator from sliding. It won't push back with "more than enough," but only just enough.

$$P - f = ma_x = 0$$

$$f = P = 170\,\text{N}$$

Friction pushes back just hard enough so that the total force is zero and the refrigerator doesn't move.

EXAMPLE

A 6 kg box sits on a 40° ramp. It is attached by a massless, frictionless rope to a 4 kg box. The coefficients of friction between the 6 kg box and the ramp are $\mu_s = 0.44$ and $\mu_k = 0.33$. What is the acceleration of each box?

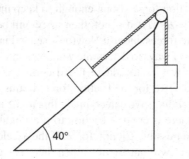

We did this problem in the last chapter except that there wasn't any friction. We found that the 6 kg box accelerates up the ramp while the 4 kg box accelerates downward. Now we repeat the process with friction.

The free body diagram is mostly the same. Each box has a gravitational force (m_1g and m_2g) and a tension force T. The left box is in contact with the ramp, so we include the unknown normal force N.

There could also be a friction force. The friction force will be in the direction to prevent (or slow) sliding of the 6 kg box on the ramp. If the 6 kg box tries to slide down the ramp the friction force will be up the ramp. If it tries to slide up the ramp the friction force will be down the ramp. If we had not already done the problem without friction, we would have to do so now in order to find out which way to put in the friction force. This is why it's best to do the friction force last when drawing free body diagrams. Fortunately we already know that, without friction, the box will slide up the ramp, so our friction force is down the ramp.

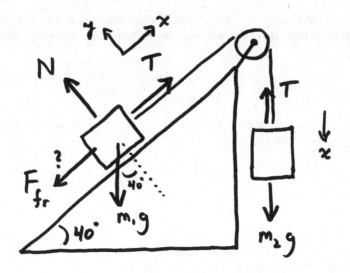

Now that we know the direction of the friction force, is it static friction or kinetic friction? Is the static friction force strong enough to keep the box from sliding? If it is then the acceleration is zero. If it is not, then there will be a kinetic friction force and we'll need to add the forces and use Newton's second law.

There are a couple of ways to determine whether static friction is strong enough to prevent sliding. One way is to add in the friction force as a variable, set the acceleration to zero, and solve for the friction force. In this way we find the friction force needed to prevent sliding. We can compare this to the maximum possible static friction force to see if there is enough friction to prevent sliding. Alternatively, we can add in the maximum possible friction force. If the acceleration is in the opposite direction as it was before (without friction) then there is enough friction to prevent sliding and there will be no acceleration. If there isn't enough friction to prevent sliding then we'll have to replace the static friction with kinetic friction and determine the acceleration. Yes, we might end up doing the calculation three times, but each iteration will be similar to the previous one. In particular, we can do the calculation once and switch the coefficients of friction.

We'll use up the ramp as the positive x direction on m_1 and down as the positive x direction on m_2, since that's the direction they'll accelerate without friction. We start by adding the forces on m_1 in the x direction.

$$F = ma$$
$$-m_1 g \sin 40° + T - \mu N = m_1 a_x$$
$$+N - m_1 g \cos 40° = m_1 a_y = 0$$
$$+m_2 g - T = m_2 a_x$$

We eliminate T by adding the first and third equations.

$$m_2 g - m_1 g \sin 40° - \mu N = (m_1 + m_2) a_x$$

We solve the second (y) equation to find the normal force

$$N = m_1 g \cos 40°$$
$$(m_1 + m_2)a_x = m_2 g - m_1 g \sin 40° - \mu m_1 g \cos 40°$$
$$(m_1 + m_2)a_x = g (m_2 - m_1 \sin 40° - \mu m_1 \cos 40°)$$

The first question is which way the boxes accelerate (we already know because we've done the problem before, but if we hadn't then this would be first). We set the coefficient of friction μ to zero and see if the acceleration is positive or negative.

$$(6 \text{ kg} + 4 \text{ kg})a_x = (9.8 \text{ m/s}^2) \big[(4 \text{ kg}) - (6 \text{ kg}) \sin 40° - (0)(6 \text{ kg}) \cos 40° \big]$$
$$a_x = +0.14 \text{ m/s}^2$$

If there is no friction, the acceleration is in the positive direction, or up the ramp. (In the last chapter it was negative, but the positive direction was down the ramp so the acceleration was up the ramp.) If we had picked the wrong direction, then we would have reversed the friction force and changed the sign of the friction force in the equation.

Is there enough static friction force to prevent sliding? Replace μ with μ_s and repeat the calculation.

$$(6 \text{ kg} + 4 \text{ kg})a_x = (9.8 \text{ m/s}^2) \big[(4 \text{ kg}) - (6 \text{ kg}) \sin 40° - (0.44)(6 \text{ kg}) \cos 40° \big]$$
$$a_x = -18.4 \text{ m/s}^2$$

If the magnitude of the static friction force were as big as it can be, the acceleration would be down the ramp. There is plenty of friction available to keep the boxes from accelerating, so the acceleration is zero.

If there were not enough friction, then the last calculation would have had the same sign as the first (been in the same direction). Then we would have needed to find the acceleration. The friction would have been kinetic, and we would have replaced the coefficient of friction μ with μ_k and crunched the numbers again.

LESSON

If we are interested in the limiting case—just as the surfaces start to slide—we can usually replace the static friction force with the maximum static friction force. If it's not the limiting case, then we need to be more careful. We might not even know the direction of the friction force.

4.2 CIRCULAR PATHS

When something goes in a circle, its velocity changes even if the speed stays the same. Since velocity is both speed and direction, an object can travel at a constant speed while the velocity changes if the direction changes. In particular, the acceleration of anything going in a circle is

$$a = \frac{v^2}{R}$$

where v is the speed and R is the radius of the circle.

If this is the acceleration, called the centripetal acceleration, then the total force must be $F = ma = mv^2/R$. This is sometimes called the centripetal force. The centripetal force is not a real force; it doesn't exist, and it never appears in a free body diagram. The "centripetal force" is the name we give to the total force, the sum of all of the forces, when something goes in a circle.

EXAMPLE

A child stands 1.2 m from the pivot of a playground merry-go-round. The child slips off if the merry-go-round turns around in less than 2.9 s. What is the coefficient of friction between the child and the merry-go-round?

We begin with the free body diagram. The child has mass so there is a gravity force mg, though we don't know the mass. She is in contact with a surface, the floor of the merry-go-round, so there is a normal force N pointing up. Because there is a normal force there could be a friction force. She isn't sliding on the floor so the friction would be static friction. She is not in contact with anything else so there are no other forces.

What is the direction of the friction force? Static friction will be whatever it needs to be in whatever direction it needs to be to prevent her from sliding on the floor. If there were no friction she would stay in place—there are no other horizontal forces. If she doesn't slide on the merry-go-round then she goes in a circle, so her acceleration is in toward the middle of the circle, toward the pivot of the merry-go-round. The friction force f must point in this direction.

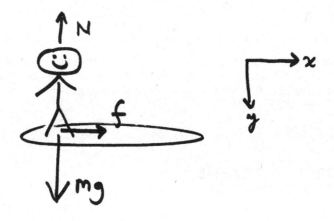

We choose the x-axis parallel to the acceleration, in toward the pivot of the merry-go-round. As the merry-go-round turns this axis will also rotate, but that's okay. We need a second axis perpendicular to the first, so the y-axis is downward (just to be different). Using Newton's second law we add the forces in the x direction.

$$F_x = ma_x$$

$$+f = m\frac{v^2}{R}$$

Since we are at the limit, the greatest speed at which she doesn't slip, we can set the static friction force equal to the coefficient of friction times the normal force.

$$\mu_s N = m\frac{v^2}{R}$$

We can find v as distance d over time t.

$$v = \frac{d}{t} = \frac{2\pi R}{t}$$

$$\mu_s N = m\left(\frac{2\pi R}{t}\right)^2 \frac{1}{R}$$

To find the normal force we add the forces in the y direction.

$$+mg - N = ma_y = 0$$

$$N = mg$$

$$\mu_s mg = \frac{4\pi^2 mR}{t^2}$$

The mass cancels—if the mass of the child doubles, her weight doubles, the normal force doubles, and the maximum static friction force doubles, but there's twice as much mass to move, so the acceleration is the same.

$$\mu_s = \frac{4\pi^2 R}{gt^2}$$

$$\mu_s = \frac{4\pi^2(1.2 \text{ m})}{(9.8 \text{ m/s}^2)(2.9 \text{ s})^2}$$

$$\mu_s = 0.57$$

EXAMPLE

A roller coaster passes over a hill at 15 m/s. What must be the radius of the curve of the hill so that the riders achieve "weightlessness" as they pass over the hill?

We start with the free body diagram. (Is this starting to get repetitive? Don't get lazy and skip this important step.) The roller coaster occupant has mass m, so there is a gravity force mg downward. He is in contact with the seat of the roller coaster, which pushes him upward with a normal force N. There could be a horizontal friction force, but since there are no other horizontal forces and no horizontal acceleration, there is no need for a friction force.

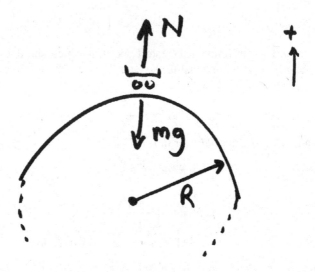

All of the forces are vertical, so we only need one axis. We'll pick upward as the positive direction.

$$F = ma$$

$$+N - mg = ma$$

As the roller coaster passes over the hill, the rider is going in a circle, or at least part of a circle. Therefore the acceleration of the rider is v^2/R toward the center of the circle, or downward.

$$+N - mg = m\left(-\frac{v^2}{R}\right)$$

$$N = mg - m\left(\frac{v^2}{R}\right)$$

$$N = m\left(g - \frac{v^2}{R}\right)$$

The term "weightlessness" is ill-named—as long as the rider is near the surface of Earth he has weight. Instead it should be called "normalforcelessness" because what we mean is when someone is not in contact with any surfaces. Gravity pulls on astronauts in orbit, but it accelerates them at the same rate as it does the spaceship around them, so that the walls don't push on them. The question is how to build the hill so that the seat doesn't push up on the rider.

But won't the rider fall? Yes, the acceleration of the rider will be downward, but as the car and rider pass over the top of the hill they go in a circle with an acceleration that's downward. Even though his speed remains constant his velocity changes because his direction changes—he was going up the hill and now he's going down the hill. If there is no normal force then it is only gravity that causes him to accelerate.

$$g - \frac{v^2}{R} = 0$$

$$R = \frac{v^2}{g} = \frac{(15 \text{ m/s})^2}{(9.8 \text{ m/s}^2)} = 23 \text{ m}$$

What if we use a smaller radius, so that N is negative? The force needed to make the rider go in a circle over the hill is greater than gravity, so additional force is needed. In some rides the handlebar yanks you down as you go over the hills, because the coaster is going too fast.

We could do the same thing for a roller coaster doing a loop. At the top of the loop the normal force is down and so is gravity. Does the coaster fall? Yes, but it needs an acceleration of gravity or more to go in the circle around the loop.

LESSON

Anything going in a circle has an acceleration of $a = v^2/R$. Everything else is the same: draw the free body diagram and set the total force equal to the acceleration.

EXAMPLE

A 1500 kg car goes around a curve of radius 200 m. The curve is banked at 11° and the coefficient of static friction between the car and the road is 0.48. What is the maximum speed of the car around the curve?

We start by drawing the free body diagram, without which all hope is lost. The car has mass so there is a gravitational force mg pulling it down. There is a normal force F_N perpendicular to the surface of the road. Because there is a normal force there could also be a friction force.

If the car goes at just the right speed the bank of the curve makes it go in a circle. This is because part of the normal force is sideways. The faster the car goes, the greater the inward force needed on the car to keep it going in a circle. The additional force comes from friction, which points down and in along the road. The friction is static friction because the tires are not sliding on the road.

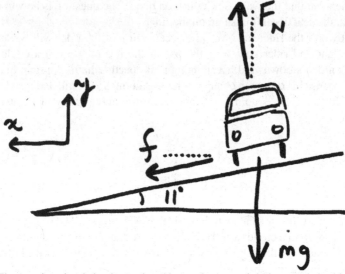

The acceleration is horizontal and inward, so we pick the x-axis in this direction. We add the forces in the x direction.

$$F_x = ma_x$$

$$+F_N \sin 11° + f \cos 11° = ma_x = m\frac{v^2}{R}$$

Because we want the maximum speed, the limiting case, the static friction force will be at its upper limit and we can replace f with μF_N.

$$+F_N \sin 11° + \mu F_N \cos 11° = m\frac{v^2}{R}$$

We want to solve this for the speed v but we don't know the normal force F_N. So we add the forces in the y direction.

$$F_y = ma_y$$

$$+F_N \cos 11° - f \sin 11° - mg = ma_y = 0$$

$$+F_N \cos 11° - \mu F_N \sin 11° = mg$$

$$F_N = \frac{mg}{(\cos 11° - \mu \sin 11°)}$$

$$\left(\frac{mg}{(\cos 11° - \mu \sin 11°)}\right)(\sin 11° + \mu \cos 11°) = m\frac{v^2}{R}$$

$$v = \sqrt{Rg\frac{(\sin 11° + \mu \cos 11°)}{(\cos 11° - \mu \sin 11°)}}$$

$$v = \sqrt{(200 \text{ m})(9.8 \text{ m/s}^2)\frac{[\sin 11° + (0.48)\cos 11°]}{[\cos 11° - (0.48)\sin 11°]}}$$

$$v = 38.2 \text{ m/s} = 85 \text{ mph}$$

An amusement park ride consists of a wide cylinder (radius = 3.8 m). People stand inside with their back against the wall. The cylinder rotates (speed of wall = 11 m/s) and then the floor is dropped away and friction holds the riders up. To increase the thrill, the ride can then be tilted by 25°. What is the minimum coefficient of friction that will hold the rider in place at the highest point in the ride?

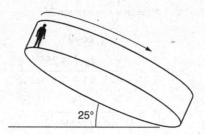

The weight of the rider W pulls him or her downward. The normal force N points perpendicularly out of the wall, or down and in. The friction f points mostly up but parallel to the wall.

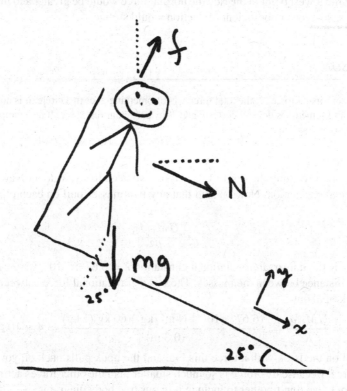

The acceleration is toward the pivot of the ride and is parallel to the normal force. We pick this as the x-axis and pick the y-axis parallel with the friction force.

$$F = ma$$

$$+N + mg \sin 25° = ma_x = m\frac{v^2}{R}$$

$$+f - mg \cos 25° = ma_y = 0$$

$$N = m\frac{v^2}{R} - mg \sin 25°$$

$$f = mg \cos 25°$$

$$\mu = \frac{f}{N} = \frac{mg \cos 25°}{(mv^2/R) - mg \sin 25°}$$

$$\mu = \frac{\cos 25°}{(v^2/Rg) - \sin 25°}$$

$$\mu = \frac{\cos 25°}{(11 \text{ m/s})^2/(3.8 \text{ m})(9.8 \text{ m/s}^2) - \sin 25°}$$

$$\mu = 0.32$$

At the lowest point in the ride the normal force would be greater and the friction force less, so a lower coefficient of friction would suffice.

LESSON

The "centripetal force" is the total force when something goes in a circle. It is not a real force and sometimes doesn't correspond to any particular real force. It never appears in a free body diagram.

In order to study the orbit of the Moon, we need to know about Newton's law of universal gravitation. Newton said that any two masses pull on each other with a force of

$$F = \frac{GM_1M_2}{R^2}$$

where G is the universal gravitational constant ($G = 6.67 \times 10^{-11}$ N m²/kg²) and R is the distance between the masses. There is a gravitational force between you and this book of about

$$F = \frac{GM_1M_2}{R^2} = \frac{(6.67 \times 10^{-11} \text{ N m}^2/\text{kg}^2)(60 \text{ kg})(1 \text{ kg})}{(0.3 \text{ m})^2} = 4 \times 10^{-8} \text{ N}$$

You pull on the book with a force this big, and the book pulls back on you with the same force. Because this force is so much smaller than the other forces acting on you or the book, we don't bother to include it in our free body diagrams.

Because the force depends on the distance, does the acceleration of gravity change as we move up and down? Yes. When we move up 3 m (about 10 ft) the gravity force becomes weaker by about one millionth, or 0.0001% weaker. At the height at which airplanes fly gravity is about 0.3% weaker than on the surface of Earth. We usually ignore this effect because it is small.

EXAMPLE

The Moon goes around Earth in a circular path of radius 3.85×10^8 m. The mass of Earth is 5.98×10^{24} kg and the mass of the Moon is 7.35×10^{22} kg. How long does it take for the Moon to orbit Earth?

The Moon has mass, so it is attracted to every other mass in the universe. In particular, it is attracted to Earth. There is nothing in contact with the Moon so there are no other forces acting on it. (This is not quite true. The Sun pulls on the Moon with a gravitational force that is about the same size as the gravity force from Earth. But the Sun also pulls on Earth so that Earth and the Moon move around the Sun together.)

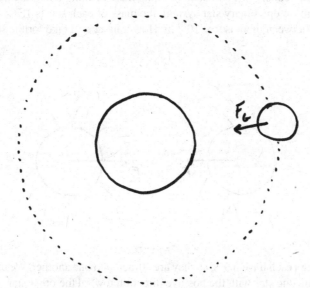

We choose positive to point toward Earth and add the force.

$$F = ma$$

$$\frac{GM_{\text{Earth}}M_{\text{Moon}}}{R^2} = M_{\text{Moon}}\frac{v^2}{R}$$

where R is the distance between Earth and the Moon, not the radius of Earth.

$$v = \sqrt{\frac{GM_{\text{Earth}}}{R}}$$

The mass of the Moon cancels. If the Moon were heavier, there would be a stronger force but also more mass to move and the same acceleration.

The time for an orbit is the distance divided by the speed.

$$t = \frac{d}{v} = \frac{2\pi R}{\sqrt{GM_{\text{Earth}}/R}} = 2\pi \sqrt{\frac{R^3}{GM_{\text{Earth}}}}$$

$$t = 2\pi \sqrt{\frac{(3.85 \times 10^8 \text{ m})^3}{(6.67 \times 10^{-11} \text{ N m}^2/\text{kg}^2)(5.98 \times 10^{24} \text{ kg})}}$$

$$t = (2.4 \times 10^6 \text{ s}) \left(\frac{1 \text{ day}}{86400 \text{ s}} \right)$$

$$t = 27.5 \text{ days}$$

EXAMPLE

A "binary star" consists of two stars of equal mass rotating about the point midway between them. In one binary star system, the mass of each star is 1.7×10^{30} kg and the distance between them is 9×10^{10} m. How long does it take for the stars to make an orbit?

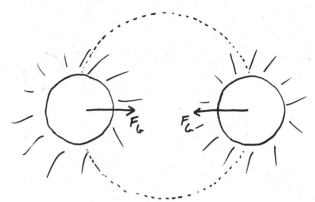

The stars each have mass, so they are attracted to one another. We use Newton's second law on one star with the positive direction toward the other star.

$$F = ma$$

$$\frac{GMM}{R^2} = M\frac{v^2}{R}$$

where R on the left is the distance between the stars and R on the right is the radius of the orbit of the stars. These are not equal! In the example with the Moon they were equal because Earth sat at the center of the Moon's circular orbit. Now they are not equal, so instead

$$\frac{GMM}{(2R)^2} = M\frac{v^2}{R}$$

where R is the radius of the orbit of the stars and $2R$ is the distance between the stars.

$$v = \sqrt{\frac{GM}{4R}}$$

$$t = \frac{d}{v} = \frac{2\pi R}{\sqrt{GM/4R}} = 4\pi \sqrt{\frac{R^3}{GM}}$$

$$t = 4\pi \sqrt{\frac{(4.5 \times 10^{10} \text{ m})^3}{(6.67 \times 10^{-11} \text{ N m}^2/\text{kg}^2)(1.7 \times 10^{30} \text{ kg})}}$$

$$t = (1.1 \times 10^7 \text{ s}) \left(\frac{1 \text{ day}}{86400 \text{ s}}\right)$$

$$t = 130 \text{ days}$$

LESSON

Be careful about what the symbols stand for. Sometimes we use the same or similar symbols in the same chapter to mean two different things. Avoid using the same symbol in the same problem to mean two different things.

CHAPTER SUMMARY

Whenever dealing with forces, always:

- Draw a free body diagram.
- Pick one axis parallel to the acceleration.
- Add the forces in each direction and set it equal to ma.
- Solve from there.

Remember:

- Friction can make something start moving since it opposes sliding, not motion.
- Static friction has a limit of $\mu_s N$, but isn't always that big.
- If something is going in a circle the acceleration is v^2/R.
- Centripetal force never goes in the free body diagram because it isn't a force, just a name.

CHAPTER **5**

ENERGY

In general, we have two ways to attack problems. The first method is to use Newton's laws, which we've just covered. The second way is to use a conservation law. If a conservation law can be used to solve a problem, it is usually the easiest way. Energy is the easiest of the conservation laws.

In a limited sense, chocolate bars are conserved. By this we mean that chocolate bars do not spontaneously appear, nor do they vanish into thin air. If there are more chocolate bars now than there were, I must have bought some. If there are less chocolate bars now than there were before, then someone has taken (and probably eaten) some. If I had six chocolate bars and now I have four, and I know that I haven't gotten any more, then I have eaten two. Just about any object is conserved in this sense.

The conservation of chocolate bars is not complete. The factory is making more all the time, and the ingredients literally grow on trees. Likewise, money is conserved in a local fashion, but since the government can print more, money is not conserved in a universal fashion.

Some things are not conserved. When two people become romantically involved, they might both get happier. This does not mean that anyone becomes less happy. Happiness is not conserved.

Since the amount of a conserved quantity cannot magically change, any change must occur by a process that we can observe. The easiest way to express this as an equation is

$$A_{\text{start}} + \Delta A = A_{\text{end}}$$

This is true for any conserved quantity. For example, money is conserved (locally). If you walk into a fast food joint with $10 and out with $6, then I know that you spent $4 even though I didn't see you do it. If I bought six chocolate bars and ate one, but there are now only three, then my kid sister must have taken two.

5.1 ENERGY AND WORK

Energy is the ability to move things, but it can do so much more. Energy is also the ability to play tunes. Wind has energy, because the wind can blow on a windmill or wind turbine, which turns a generator and creates electricity, which powers my CD player and amplifier, playing tunes (playing tunes is the purpose of physics in the first place). This view of energy works because sound is a form of energy, so anything that can be turned into sound is also a form of energy. Energy appears in many forms and can be transformed between forms, though the transformation may be incomplete, inefficient, or expensive.

We are mostly concerned with the forms of energy and work called mechanical energy, which includes kinetic energy and potential energy. Kinetic energy is the energy that a moving object has, and is equal to $KE = \frac{1}{2}mv^2$. Rotating objects also have kinetic energy, and I'll introduce that when we get there. The unit of energy and of work is the joule, which is equal to a newton times a meter.

$$1\,J = 1\,N\,m = 1\,kg\,m^2/s^2$$

Work is the change in the mechanical energy. Mathematically, work is the force times the displacement, or

$$W = \boldsymbol{F} \cdot \boldsymbol{d}$$

The "dot product" means that only the part of the vector $\boldsymbol{F}$ that is parallel to the vector $\boldsymbol{d}$ matters. For example, when the force on a car is in the same direction that it is moving (forward), the car speeds up and the energy increases. Because the force and the displacement are in the same direction, the work done is positive, which matches the increase in the energy of the car. If the force on a car is in the opposite direction that it is moving, the car slows down and the energy decreases. Because the force and the displacement are in opposite directions, the work done is negative, which matches the decrease in the energy of the car. The work is usually calculated using

$$W = Fd\cos\theta$$

where θ is the angle between the force F and the displacement d.

Consider an example involving negative work. If a car is moving and we want to stop it, we must take energy away from the car, so we must do negative work on the car. This energy is then turned into some other form of energy. We could, for example, tie a rope to the back of the car and wrap the other end of the rope around the axle of a generator, so that as the tension in the rope slows the car it also turns the generator, making electricity, powering our boombox, and creating tunes. When we calculate the work, we use an angle of 180°, and the cosine of 180° is -1, so we get a negative work.

EXAMPLE

A 6 kg box is dropped from a height of 3 m. How fast is it going when it hits the floor?

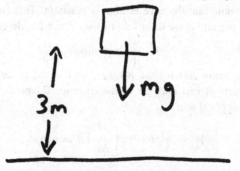

We could do this using constant acceleration techniques. However, in order to demonstrate energy, we'll use conservation of energy.

$$E_{before} + W = E_{after}$$

$$\frac{1}{2}mv_0^2 + W = \frac{1}{2}mv^2$$

The only force acting on the box is gravity, so the work done is the force of gravity times the distance traveled.

$$\frac{1}{2}m(0)^2 + (mg)d = \frac{1}{2}mv^2$$

$$gd = \frac{1}{2}v^2$$

$$v = \sqrt{2gd} = \sqrt{2(9.8 \text{ m/s}^2)(3 \text{ m})}$$

$$v = 7.7 \text{ m/s}$$

EXAMPLE

How much work does the engine need to do to accelerate a 1500 kg car from rest to 35 mph?

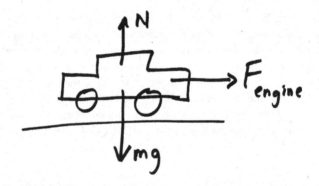

First let's assume that the acceleration is constant. This isn't necessarily true, but it's probably close and if we didn't do this we'd need to do calculus.

$$W = Fd = mad$$

We don't care how much time it takes, so $v^2 - v_0^2 = 2a \, \Delta x$, where $v_0 = 0$ because the car starts at rest and $\Delta x = $ the distance traveled = the distance over which the force is applied = d.

$$W = mad = m\left(\frac{v^2}{2d}\right)d = \frac{1}{2}mv^2$$

$$W = \frac{1}{2}(1500\ \text{kg})\left[(35\ \text{mi/h})\left(\frac{1610\ \text{m}}{1\ \text{mi}}\right)\left(\frac{1\ \text{h}}{3600\ \text{s}}\right)\right]^2$$

$$W = 1.84 \times 10^5\ \text{J} \quad \text{or} \quad 184\ \text{kJ}$$

This is about the energy in 0.004 gallons of gasoline. Since gas engines are not perfectly efficient, it takes about 1% of a gallon of gas to accelerate a car up to city speed. Alternatively, we can use conservation of energy.

$$E_0 + W = E$$

$$KE_0 + W = KE$$

$$\frac{1}{2}m(0)^2 + W = \frac{1}{2}mv^2$$

$$W = \frac{1}{2}mv^2$$

This is the same equation as above, but it doesn't require that the acceleration be constant. It also avoids the need to do tricky things with equations.

EXAMPLE

A 12 kg box is pulled 3.0 m across the floor by a rope. The rope has a tension of 60 N and is 37° above the horizontal. The coefficient of friction between the box and the floor is 0.23. What is the speed of the box after being pulled?

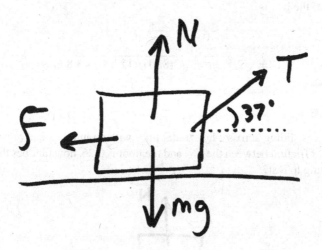

We can use conservation of energy to find the speed of the box.

$$KE + W = KE'$$

$$\frac{1}{2}m(0)^2 + W = \frac{1}{2}m(v')^2$$

The work done by the rope is

$$W_T = Fd \cos \theta = (60 \text{ N})(3 \text{ m}) \cos 37° = 144 \text{ J}$$

The work done by gravity is

$$W_{mg} = Fd \cos \theta = (12 \text{ kg})(9.8 \text{ m/s}^2)(3 \text{ m}) \cos 90° = 0$$

The work done by the normal force is

$$W_N = Fd \cos \theta = (N)(3 \text{ m}) \cos 90° = 0$$

The work done by friction is

$$W_F = Fd \cos \theta = \mu N d \cos 180°$$

We find the normal force as we did earlier.

$$+N + T \sin 37° - mg = ma_y = 0$$
$$N = mg - T \sin 37°$$
$$N = (12 \text{ kg})(9.8 \text{ m/s}^2) - (60 \text{ N}) \sin 37°$$
$$N = 81.5 \text{ N}$$
$$W_F = (0.23)(81.5 \text{ N})(3 \text{ m}) \cos 180° = -56 \text{ J}$$

The work done by all forces is

$$144 \text{ J} + 0 + 0 + (-56 \text{ J}) = 88 \text{ J}$$

The speed of the box is

$$0 + W = \frac{1}{2}m(v')^2$$
$$v' = \sqrt{2W/m} = \sqrt{2(88 \text{ J})/(12 \text{ kg})} = 3.8 \text{ m/s}$$

EXAMPLE

A 14 kg box is sliding across a horizontal floor with an initial speed of 6 m/s. If the coefficient of friction between the box and the floor is 0.26, how far does the box slide before coming to rest?

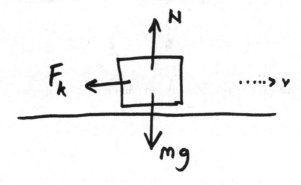

Using conservation of energy,

$$E_{\text{before}} + W = E_{\text{after}}$$

$$\frac{1}{2}mv_0^2 + W = 0$$

The forces acting on the box are gravity, the normal force from the floor, and the friction force from the floor. Gravity and the normal force from the floor are vertical forces, and the displacement is horizontal, so these forces are perpendicular to the motion and don't do any work. The only force that does any work is the friction force, which is in the opposite direction as the motion and does negative work. The force of friction is $\mu N = \mu mg$, since there is no vertical acceleration and the vertical forces add to zero.

$$\frac{1}{2}mv_0^2 - (\mu mg)d = 0$$

$$\frac{1}{2}v_0^2 - \mu gd = 0$$

$$d = \frac{v_0^2}{2\mu g} = \frac{(6 \text{ m/s})^2}{2(0.26)(9.8 \text{ m/s}^2)} = 7.1 \text{ m}$$

5.2 POTENTIAL ENERGY

Some types of work can be expressed as potential energy. Potential energy is work done in the past that could be turned into work again. Gravity, for example, has potential energy and a compressed spring has potential energy. On the other hand, the work done by friction cannot be expressed as a potential energy.

What makes gravity different is that if you lift a box above your head, the work done by gravity during the process does not depend on the path taken; the work done by gravity is the same whether you lift the box straight up, push it up a ramp, or follow a curved path. A quick check is to (hypothetically) move the object back and forth or in a circle, back to where it started; if the work done by a force *must* be zero, then it might be possible to use potential energy for that force. If we move a box back and forth across the floor, the force of friction is always against the motion and the work done by the friction force is negative, not zero, so friction cannot be expressed using a potential energy.

The reason why we use potential energy is because the energy doesn't depend on the path but only on the position. This makes *potential energy easier to calculate than work.* If we use potential energy to express the work done by a force, then we don't include the work by this force when we calculate the work. There are three forces that can commonly be expressed with potential energy: gravity, springs, and electricity. Electricity comes later, but the first two are covered here.

The potential energy from gravity is

$$PE_{\text{gravity}} = mgh$$

In other words, the work we do lifting something is the force mg times the distance h. What is the height? We can choose *anywhere* to be zero height, from which all heights are measured. Every time we use conservation of energy in an equation, the potential energy before and the potential energy after both show up, so that only the change in potential energy matters. Therefore, we can choose any spot to be zero height, but once we choose we must stick with that spot for the whole problem (like a coordinate axis).

EXAMPLE

How much work does it take to carry a 30 pound pack 10 miles over level ground? We can either calculate the work directly or use conservation of energy.

First let's use conservation of energy.

$$KE + PE + W = KE' + PE'$$

$$\frac{1}{2}m(0)^2 + mgh + W = \frac{1}{2}m(0)^2 + mgh'$$

The height is the same before and after, so it cancels and the work is zero. Since the speed of the pack has not increased, and its height didn't change, the energy didn't change, so the work is zero. It's true that we must have done a little work at the beginning to get the pack moving, but we did a little negative work at the end to stop it, and they cancel.

Now let's do the same problem by finding the work directly. The force we apply to the pack is upward and equal in magnitude to gravity, so that the total force on the pack is zero. With no total force, the pack has no acceleration and cruises the 10 miles at a constant velocity. The force we apply is upward, which is perpendicular to the sideways motion, so the work we do is zero.

EXAMPLE

If an 11 kg box is dropped from a height of 1.7 m, how fast will it be going when it hits the ground?

We could solve this using constant acceleration.

$$v^2 - v_0^2 = 2a \ \Delta x$$

$$v^2 - 0 = 2(9.8 \text{ m/s}^2)(1.7 \text{ m})$$

$$v = 5.8 \text{ m/s}$$

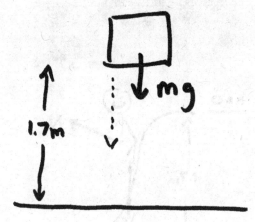

We want to use conservation of energy (as a demonstration). We choose the floor as zero height.

$$KE_{before} + PE_{before} + W = KE_{after} + PE_{after}$$

$$\frac{1}{2}m(0)^2 + mgh + W = \frac{1}{2}mv^2 + mg(0)$$

The only force acting on the box as it falls is gravity. Because we have expressed the work done by gravity as the potential energy of gravity, we don't include that force when calculating work. Since there are no other forces, there is no work. (Alternatively, we could have omitted the potential energy and included the work done by gravity, $W = Fd = mgh$.)

$$mgh = \frac{1}{2}mv^2$$

$$(9.8 \text{ m/s}^2)(1.7 \text{ m}) = \frac{1}{2}v^2$$

$$v = 5.8 \text{ m/s}$$

EXAMPLE

If a 11 kg child slides down a frictionless slide from a height of 1.7 m, how fast will she be going when she reaches the ground?

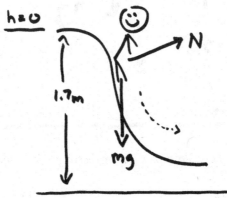

Because the slide is not straight, the acceleration is not constant, and we can't use constant acceleration techniques. Instead we use conservation of energy. For variety, we choose the top of the ramp as zero height:

$$KE_i + PE_i + W = KE_f + PE_f$$

$$\frac{1}{2}m(0)^2 + mg(0) + W = \frac{1}{2}m(v_f)^2 + mgh_f$$

where $h_f = -1.7$ m.

The only forces acting on the child as she slides are gravity and the normal force. Because we have expressed the work done by gravity as the potential energy of gravity, we don't include that force when calculating work. The normal force is always perpendicular to the surface, and the motion is always parallel to the surface. These are perpendicular, so the work done by the normal force is zero.

$$0 = \frac{1}{2}m(v_f)^2 + mgh_f$$

$$-\frac{1}{2}(v_f)^2 = (9.8 \text{ m/s}^2)(-1.7 \text{ m})$$

$$v_f = 5.8 \text{ m/s}$$

This is the same result as in the previous example. The work done by gravity is the same, because we can express it as a potential energy that depends only on the end points. The work done by other forces (the normal force) is zero, so the kinetic energy is the same and the speed is the same.

LESSON

The advantage of using a conservation law is that it works even when the acceleration isn't constant.

EXAMPLE

A 300 kg piano is lifted two stories (7.2 m) by a freight elevator. How much work does the elevator do on the piano?

We could find the work directly by multiplying the force times the distance. To do this we must assume that the elevator lifts the piano at a constant speed, so that the normal force is equal to the weight. Instead let's try conservation of energy, choosing the starting position as zero height.

$$KE_{\text{before}} + PE_{\text{before}} + W = KE_{\text{after}} + PE_{\text{after}}$$
$$\frac{1}{2}m(0)^2 + mg(0) + W = \frac{1}{2}m(0)^2 + mgh$$
$$W = mgh = (300 \text{ kg})(9.8 \text{ m/s}^2)(7.2 \text{ m})$$
$$W = 2.1 \times 10^4 \text{ J} \quad \text{or} \quad 21 \text{ kJ}$$

This work is done by the normal force. The normal force pushes up while the motion is also up. The normal force is increasing the speed of the piano while gravity is decreasing the speed of the piano.

LESSON

If the surface doesn't move, the normal force doesn't do any work.

Springs are another force that can be treated using potential energy. The force from a spring is $F = -kx$, where k is the "spring constant" and x is how much the spring has been stretched or compressed. As the spring is compressed, the force from the spring changes and the acceleration changes. The potential energy is the work done compressing the spring, which is

$$PE_{\text{spring}} = \frac{1}{2}kx^2$$

Why the 1/2? The work is the distance x times the average force $\frac{1}{2}kx$.

EXAMPLE

A horizontal spring ($k = 1000$ N/m) is compressed 6 cm and a 3 kg toy car is placed in front of it. When the spring is released, what speed does the car obtain?

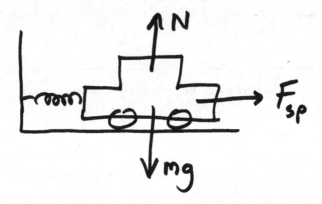

Doing this problem with forces would involve finding the acceleration, which changes as the compression of the spring changes. This would lead to a differential equation and a very difficult problem. Likewise, finding the work involves finding a force that is changing over time and position, making for a difficult problem.

On the other hand, the potential energy of a compressed spring is $PE = \frac{1}{2}kx^2$. Using conservation of energy,

$$KE + PE + W = KE' + PE'$$

$$\frac{1}{2}m(0)^2 + \left(\frac{1}{2}kx^2 + mgh\right) + W = \frac{1}{2}m(v')^2 + \left(\frac{1}{2}k(0)^2 + mgh'\right)$$

The work does not include the work done by gravity or the spring, because we already included them by using potential energy. The gravitational potential energy cancels because $h = h'$. The only other force is the normal force from the floor, but the normal force is vertical and the motion is horizontal, so the normal force doesn't do any work.

$$\frac{1}{2}kx^2 = \frac{1}{2}m(v')^2$$

$$v = \sqrt{\frac{kx^2}{m}}$$

$$v = \sqrt{\frac{(1000 \text{ N/m})(0.06 \text{ m})^2}{3 \text{ kg}}}$$

$$v = 1.1 \text{ m/s}$$

EXAMPLE

A 4 kg box is dropped from a height of 2 m onto a vertically oriented spring ($k = 1200$ N/m). How far will the spring compress before the box comes momentarily to rest?

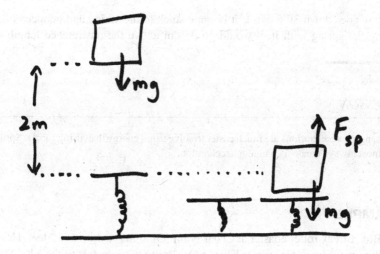

The spot at which the box stops (momentarily) is not when the spring and gravity forces are equal. When the forces are equal the acceleration is zero, but for the box to stop the acceleration must be upward, opposite the velocity.

As in the last example, the force on the box changes with time as the spring is compressed, making it very difficult to do this problem with forces. Instead, we use conservation of energy, choosing the top of the uncompressed spring as zero height,

$$KE_{before} + PE_{before} + W = KE_{after} + PE_{after}$$

$$KE_A + PE_A + W = KE_C + PE_C$$

$$\frac{1}{2}m(0)^2 + \left(\frac{1}{2}k(0)^2 + mgh\right) + W = \frac{1}{2}m(0)^2 + \left(\frac{1}{2}kx^2 + mg(-x)\right)$$

where x is how far the spring is compressed, at which point the height of the box is x below zero. The work is zero, since the only forces are gravity and the spring and both have been dealt with using potential energy.

$$mgh = \frac{1}{2}kx^2 - mgx$$

$$\frac{1}{2}kx^2 - mgx - mgh = 0$$

$$x = \frac{mg \pm \sqrt{(-mg)^2 - 4(\frac{1}{2}k)(-mgh)}}{2(\frac{1}{2}k)}$$

$$x = \frac{(4 \text{ kg})(9.8 \text{ m/s}^2) \pm \sqrt{[(4 \text{ kg})(9.8 \text{ m/s}^2)]^2 + 2(1200 \text{ N/m})(4 \text{ kg})(9.8 \text{ m/s}^2)(2 \text{ m})}}{(1200 \text{ N/m})}$$

$$x = \frac{(39.2 \text{ kg m/s}^2) \pm (435.5 \text{ kg m/s}^2)}{(1200 \text{ N/m})}$$

$$x = 0.396 \text{ m} \quad \text{or} \quad -0.330 \text{ m}$$

The box goes down 39.6 cm. If it became stuck to the spring and bounced back up, pulling the spring with it, it would go 33 cm above the unstretched length of the spring.

LESSON

Changing accelerations are much easier to solve using energy than using forces. Springs almost always create a changing acceleration.

EXAMPLE

The Blue Streak roller coaster at Cedar Point (Ohio) has a drop of 72 feet. How high of a drop would a roller coaster have to have to go twice as fast as the Blue Streak?

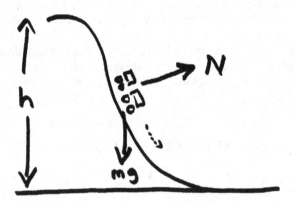

The acceleration is not constant, so we use conservation of energy and choose the bottom of the hill as zero height.

$$KE_{before} + PE_{before} + W = KE_{after} + PE_{after}$$
$$\frac{1}{2}m(0)^2 + mgh + W = \frac{1}{2}mv^2 + mg(0)$$

The work done by gravity is taken care of with potential energy. The work done by the normal force is zero because the normal force is perpendicular to the motion. The work done by friction is difficult to calculate, but small enough that we will ignore it here.

$$mgh + 0 = \frac{1}{2}mv^2$$
$$gh = \frac{1}{2}v^2$$

where v is the speed of the Blue Streak at the bottom of the hill.

The question here is not how fast the Blue Streak is going (v), but what height (h) we would need to go twice as fast. Using h' as the height of the new coaster and v' as the speed of the new coaster,

$$gh' = \frac{1}{2}v'^2$$

$$h' = \frac{1}{2}\frac{v'^2}{g} = \frac{1}{2}\frac{(2v)^2}{g} = 4\left(\frac{1}{2}\frac{v^2}{g}\right)$$

At this point we substitute in the result from the old coaster.

$$h' = 4\frac{1}{2}\frac{v^2}{g} = 4\frac{gh}{g} = 4h$$

The new coaster must be four times higher (almost 300 ft) to go twice as fast.

Not far from the Blue Streak is the newer, higher, faster Millennium Force. The Millennium Force is just over four times higher than the Blue Streak and about twice as fast.

EXAMPLE

A spring is compressed near the bottom of a ramp. If the mass of the block is 0.42 kg, the spring of spring constant 1800 N/m is compressed by 0.060 m, the angle is 25°, and the coefficient of kinetic friction is 0.18, how far up the ramp does the block go?

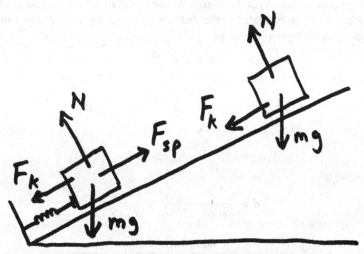

Because the force of the spring is not constant, the acceleration is not constant and using forces looks unpromising. Instead we try conservation of energy, choosing the starting position of the block as zero height.

$$KE + PE + W = KE' + PE'$$

$$\frac{1}{2}m(0)^2 + \left(mg(0) + \frac{1}{2}kx^2\right) + W = \frac{1}{2}m(0)^2 + \left(mgh' + \frac{1}{2}k(0)^2\right)$$

$$\frac{1}{2}kx^2 + W = mgh'$$

The work done by the spring and by gravity are handled with potential energy. The normal force is perpendicular to the motion, so the work done by the normal force is zero. The friction force does negative work, slowing the block, because the friction force and the motion are in opposite directions. We find the normal force as we did earlier. The work done by friction is

$$W_F = F_F d \cos\theta = (\mu N) d \cos 180°$$

The normal force is

$$+N - mg\cos 25° = ma_y = 0$$

$$N = mg\cos 25°$$

$$\frac{1}{2}kx^2 - \mu \underbrace{mg\cos 25°}_{N}\, d = mgh'$$

In that equation, d is the distance along the ramp, h is the vertical height, and x is the original compression of the spring. We are trying to solve for d and x is known. The vertical height h is related to the distance d, in that the further along the ramp the block goes the higher it goes, by

$$h' = d\sin 25°$$

Therefore

$$\frac{1}{2}kx^2 - \mu mgd\cos 25° = mg(d\sin 25°)$$

$$d\,(mg\sin 25° + \mu mg\cos 25°) = \frac{1}{2}kx^2$$

$$d = \frac{kx^2}{2mg(\sin 25° + \mu\cos 25°)}$$

$$d = \frac{(1800\ \text{N/m})(0.060\ \text{m})^2}{2(0.42\ \text{kg})(9.8\ \text{m/s}^2)(\sin 25° + 0.18\cos 25°)}$$

$$d = 1.3\ \text{m}$$

5.3 USING CONSERVATION OF ENERGY

There are many problems that can be solved using conservation of energy, but there are many problems that cannot. In this section we'll explore some of the reasons why problems can't be solved using conservation of energy.

EXAMPLE

An Atwood's machine consists of blocks of mass 15 kg and 5 kg hanging over a massless, frictionless pulley. How fast will the blocks be going after they have each moved 3 m, and how long will it take for this to happen?

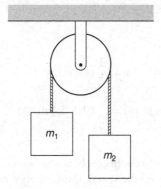

We could use forces to determine the acceleration, which would be a constant, and then use constant acceleration techniques to find the velocity. Instead let's try this with conservation of energy, taking the starting point of each block as zero height for that block.

$$KE_{before} + PE_{before} + W = KE_{after} + PE_{after}$$

$$\tfrac{1}{2}(m_1 + m_2)(0)^2 + 0 + W = \tfrac{1}{2}(m_1 + m_2)v^2 + [m_1 g(h) + m_2 g(-h)]$$

where $h = 3$ m. The work done by gravity has been handled by potential energy, so we only need to find the work done by the tension forces.

The work done by the left tension on the 5 kg block is positive, because the force is in the same direction as the motion and because the tension increases the speed of the block. The work done by the right tension on the 15 kg block is negative, because the force is in the opposite direction as the motion and because the tension holds back the block, decreasing its speed. Whatever the positive work the left tension does on the 5 kg block, the right tension does the same amount of negative work on the 15 kg block. The forces are the same and the distances are the same, so the works are the same but opposite. The total work done on the system (other than by gravity) is zero.

$$0 = \frac{1}{2}(m_1 + m_2)v^2 + gh\,(m_1 - m_2)$$

$$0 = \frac{1}{2}(20\text{ kg})v^2 + (9.8\text{ m/s}^2)(3\text{ m})(-10\text{ kg})$$

$$v = \sqrt{(9.8\text{ m/s}^2)(3\text{ m})} = 5.4\text{ m/s}$$

If we assume that the acceleration is constant then we can find the time.

$$\Delta x = 3\text{ m}$$
$$v_0 = 0$$
$$v = 5.4\text{ m/s}$$
$$a =$$
$$t = ?$$

$$\Delta x = \frac{1}{2}(v_0 + v)t$$

$$(3\text{ m}) = \frac{1}{2}(0 + 5.4\text{ m/s})t$$

$$t = 1.1\text{ s}$$

If we had used forces to do this problem, we would have found the acceleration first and then used that to find the speed. Using energy we do this backward, finding the speed and then finding the acceleration. To do this we have to know that the acceleration is constant, but there is nothing in our solution that says that it must be.

LESSON

Conservation laws work well even when the acceleration is not constant, so they typically can't be used to find the acceleration.

EXAMPLE

A compact car (1200 kg) traveling at 25 m/s collides with a stationary truck (8000 kg) and the two stick together. What is the speed of the pair of vehicles just after the collision?

This would be difficult to solve using forces. The force of the car on the truck is the same as but opposite to the force of the truck on the car. However, we don't know either the force or the time over which the force was applied. Instead we try conservation of energy.

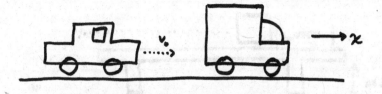

$$KE_{\text{before}} + PE_{\text{before}} + W = KE_{\text{after}} + PE_{\text{after}}$$

$$\left[\tfrac{1}{2}m_{\text{car}}v_0^2 + \tfrac{1}{2}m_{\text{truck}}(0)^2\right] + \left[m_{\text{car}}g(0) + m_{\text{truck}}g(0)\right] + W$$

$$= \tfrac{1}{2}\left(m_{\text{car}} + m_{\text{truck}}\right)v^2 + \left[m_{\text{car}}g(0) + m_{\text{truck}}g(0)\right]$$

$$\left(\tfrac{1}{2}m_{\text{car}}v_0^2\right) + W = \tfrac{1}{2}(m_{\text{car}} + m_{\text{truck}})v^2$$

The work done by gravity has been handled by potential energy, and the normal force is perpendicular to the motion ($W_{\text{normal}} = 0$), so we only need to find the work done by forces the car and truck put on each other.

The forces the car and truck put on each other are the same, but the distances are not necessarily the same. During the collision, the car will be dented more than the truck (this may not be obvious now). Since the distances are not the same, the works are not equal and opposite as they were in the last example. The total work is not zero, but still unknown. Without knowing the work, we cannot solve this problem using conservation of energy.

The total energy in this collision is conserved, but some of the mechanical energy is turned into thermal energy—the car and the truck are slightly warmer after the collision than before. If the amount of heat generated is not known, then conservation of energy, while still true, can't be used to solve the problem. (We will solve this in the next chapter.)

LESSON

When mechanical energy gets turned into heat (other than by friction), conservation of energy typically can't be used to solve the problem.

EXAMPLE

A spring-loaded gun sits atop a table as shown. The table is 1.2 m high, the "bullet" is a frictionless block of mass 0.40 kg, the spring constant of the gun is 1500 N/m, and the spring is compressed 0.14 m. How far past the edge of the table will the "bullet" hit the ground?

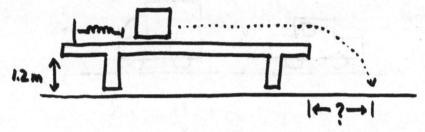

1.2 m

We could try to solve this using forces; finding how far it goes as it flies through the air sounds like something we did using constant acceleration. However, the force of the spring as it accelerates the object is not constant, and finding it using forces would be difficult.

Instead we try conservation of energy, choosing the top of the table as zero height.

$$KE_{before} + PE_{before} + W = KE_{after} + PE_{after}$$

We can account for the spring and gravity forces using potential energy, and the normal force doesn't do any work (it's perpendicular to the motion). But there's nothing in the equation that connects to a horizontal distance. Even the work done by gravity is exerted over a vertical distance. We can't solve this problem using energy.

We've seen that constant acceleration works well for things flying through the air, and that energy works well for springs (which have a nonconstant force). Let's use each technique where it works well, using energy to find the speed of the object as it leaves the table and constant acceleration to find how far it goes.

$$KE_{before} + PE_{before} + W = KE_{after} + PE_{after}$$

$$\frac{1}{2}m(0)^2 + \left(mg(0) + \frac{1}{2}kx^2 \right) + 0 = \frac{1}{2}mv^2 + \left(mg(0) + \frac{1}{2}k(0)^2 \right)$$

$$\frac{1}{2}kx^2 = \frac{1}{2}mv^2$$

$$v = \sqrt{\frac{kx^2}{m}}$$

$$v = \sqrt{\frac{(1500 \text{ N/m})(0.14 \text{ m})^2}{0.40 \text{ kg}}} = 8.6 \text{ m/s}$$

We find the time to fall to the floor using constant acceleration. Choosing down as the positive direction,

$$\Delta y = v_{y0}t + \frac{1}{2}a_y t^2$$

$$1.2 \text{ m} = (0)t + \frac{1}{2}(9.8 \text{ m/s}^2)t^2$$

$$t = \sqrt{\frac{2(1.2 \text{ m})}{(9.8 \text{ m/s}^2)}} = 0.49 \text{ s}$$

There is no horizontal acceleration, and the horizontal distance is

$$\Delta x = v_x t = (8.6 \text{ m/s})(0.49 \text{ s}) = 4.2 \text{ m}$$

LESSON

It's possible to use multiple techniques, such as forces and conservation of energy, to do parts of the same problem. Use each technique where it works best.

EXAMPLE*

A 11 kg child slides down a frictionless slide that is in the shape of a quarter-circle with radius of 1.7 m. How long will it take her to reach the ground? How long will it take if the coefficient of kinetic friction between the child and the slide is 0.14?

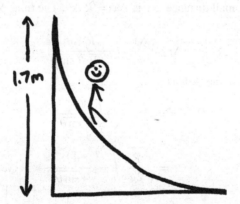

Not only is the velocity not constant, the acceleration is not constant either. We solve this using calculus, where the time Δt for each small step Δx is $\Delta t = \Delta x / v$. We use conservation of energy to find the speed v, choosing the top of the slide as zero height.

$$KE_{\text{before}} + PE_{\text{before}} + W = KE_{\text{after}} + PE_{\text{after}}$$

*Example uses calculus.

$$\frac{1}{2}m(0)^2 + mg(0) + (0) = \frac{1}{2}mv^2 + mg(-R\sin\theta)$$

$$\frac{1}{2}v^2 = gR\sin\theta$$

$$v = \sqrt{2gR\sin\theta}$$

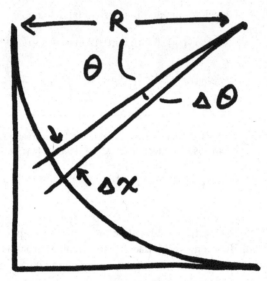

The length of the small distance Δx is $\Delta x = R\,\Delta\theta$. The time Δt for the small step Δx is

$$\Delta t = \frac{\Delta x}{v} = \frac{R\,\Delta\theta}{\sqrt{2gR\sin\theta}}$$

The time to go down the slide is

$$t = \int_0^{\pi/2} \frac{R\,d\theta}{\sqrt{2gR\sin\theta}}$$

$$t = \sqrt{\frac{R}{2g}} \int_0^{\pi/2} \frac{d\theta}{\sqrt{\sin\theta}}$$

$$t \approx \sqrt{\frac{R}{2g}} \left(\frac{5}{3}\right)$$

where the integral has been done numerically.

$$t \approx \sqrt{\frac{(1.7\,\text{m})}{2(9.8\,\text{m/s}^2)}} \left(\frac{5}{3}\right)$$

$$t \approx 0.49\,\text{s}$$

To do this with friction, we would need to find the work done by friction. This means we would need to find the normal force at each point along the journey, which depends on the acceleration, which depends on the speed, which depends on the friction force at all previous points along the journey. Solving this would require nasty differential equations.

CHAPTER SUMMARY

- Energy is conserved, so

$$\underbrace{\frac{1}{2}mv^2}_{KE_{before}} + \underbrace{PE_{before}}_{mgh + \frac{1}{2}kx^2} + \overbrace{W}^{Fd\ \cos\ \theta} = KE_{after} + PE_{after}$$

- If you use potential energy for a force, then don't include the work done by that force.
- Forces that act perpendicular to the motion don't do any work.

CHAPTER 6

MOMENTUM

Just as energy is conserved, so is momentum. When a large truck collides head-on with a small car, we expect to see the car bounce backward while the truck continues on. This is not because the force of the truck on the car is larger than the force of the car on the truck—the forces are the same. Instead, this is because the truck has more momentum than the car.

Conservation of momentum looks very similar to conservation of energy. The equations have the same general form and the techniques are the same. Conservation of momentum works particularly well for collisions.

6.1 MOMENTUM AND IMPULSE

Momentum is the product of mass and velocity:

$$P = mv$$

Impulse is the name given to the change in the momentum ΔP, which thus has the same units as momentum (kg m/s or N s). The basic equation is

$$P_{\text{before}} + \Delta P = P_{\text{after}} \quad \text{or} \quad P + \Delta P = P'$$

$$\Delta P = Ft$$

One difference between momentum and energy is that momentum is a vector, so it has direction.

EXAMPLE

A 0.2 kg rubber ball is thrown at a brick wall at a speed of 4 m/s and bounces back at a speed of 4 m/s; what is the work done by the wall and the impulse exerted by the wall on the rubber ball?

Since the speed of the ball is unchanged, the kinetic energy is unchanged, the energy is unchanged, and the work is zero.

$$KE_{\text{before}} + PE_{\text{before}} + W = KE_{\text{after}} + PE_{\text{after}}$$

$$\frac{1}{2}(0.2 \text{ kg})(4 \text{ m/s})^2 + mgh + W = \frac{1}{2}(0.2 \text{ kg})(-4 \text{ m/s})^2 + mgh$$

$$1.6 \text{ J} + W = 1.6 \text{ J}$$

$$W = 0$$

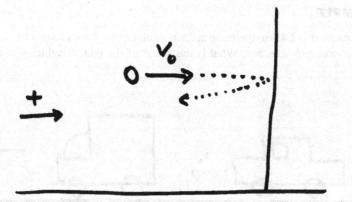

The magnitude of the momentum is the same before and after, but the directions are different, so the momentums are not the same and the impulse is not zero.

$$P_{before} + \Delta P = P_{after}$$
$$mv_{before} + \Delta P = mv_{after}$$
$$(0.2 \text{ kg})(4 \text{ m/s})\,\hat{i} + \Delta P = (0.2 \text{ kg})(-4 \text{ m/s})\,\hat{i}$$

Here $\hat{i}$ is a unit vector in the x direction—a vector of magnitude 1 and no units. The impulse is

$$0.8 \text{ kg m/s }\hat{i} + \Delta P = -0.8 \text{ kg m/s }\hat{i}$$
$$\Delta P = -1.6 \text{ kg m/s }\hat{i}$$

As the ball bounces off of the wall the displacement is zero. The wall does negative work on the ball as it slows down, but as the ball bounces back and speeds up the wall does positive work. Since the energy of the ball is unchanged the total work the wall does on the ball is zero.

The impulse (change in momentum) the wall gives to the ball is the force times the time. Since the force is in the same direction (negative x) when the ball is both slowing and rebounding, and the time is positive, the impulse is in the negative x direction, outward from the wall.

LESSON

Momentum is a vector and has direction. A change in direction might not change the energy but it must change the momentum.

6.2 MOMENTUM AND COLLISIONS

Conservation of momentum is usually the best way to handle collisions. This is because the forces that the two objects apply to one another cancel out.

EXAMPLE

A compact car (1200 kg) traveling at 25 m/s collides with a stationary truck (8000 kg) and the two stick together. What is the speed of the pair of vehicles just after the collision?

We tried to apply conservation of energy to this problem in the last chapter but were unable to get an answer. Now we try conservation of momentum. We treat the two vehicles together, so that P is the momentum of both objects.

$$P + \Delta P = P'$$

$$m_{car}v_{car} + m_{truck}v_{truck} + \Delta P = m_{car}v'_{car} + m_{truck}v'_{truck}$$

The initial velocity of the truck v_{truck} is zero, and the two vehicles stick together so $v'_{car} = v'_{truck}$

$$m_{car}(25 \text{ m/s}) + (F_{car \text{ on } truck}t - F_{car \text{ on } truck}t) = m_{pair}v'_{pair}$$

Here t is the time over which the collision takes place. We drop the vector notation and look at only the momentum (and forces) in the $+x$ direction.

We don't know the force of the car on the truck, nor the force of the truck on the car. What we do know is that those forces are the same. Those terms in the equation must add to zero.

$$m_{car}(25 \text{ m/s}) + (0t) = m_{pair}v'_{pair}$$

$$m_{car}(25 \text{ m/s}) = m_{pair}v'_{pair}$$

$$v'_{pair} = \frac{m_{car}(25 \text{ m/s})}{m_{pair}}$$

$$v'_{pair} = \frac{(1200 \text{ kg}) (25 \text{ m/s})}{(1200 \text{ kg} + 8000 \text{ kg})}$$

$$v'_{pair} = 3.3 \text{ m/s}$$

EXAMPLE

A 60 kg girl on roller blades throws a 0.60 kg ball, giving it a speed of 10 m/s. What is her (backward) speed after throwing the ball?

We don't know the forces that she applies to the ball or that the ball exerts on her, so using forces doesn't look good. We could try to use conservation of energy, but we don't know the work she does on the ball. Instead we try conservation of momentum.

$$P + \Delta P = P'$$
$$(m_{girl}v_{girl} + m_{ball}v_{ball}) + (F_{girl\ on\ ball}t - F_{ball\ on\ girl}t) = (m_{girl}v'_{girl} + m_{ball}v'_{ball})$$

As in the last example, the force of the girl on the ball and the force of the ball on the girl are the same but in opposite directions. Those two terms add to zero. Neither the girl nor the ball are moving before she throws it, so the two leftmost terms are zero.

$$0 = (m_{girl}v'_{girl} + m_{ball}v'_{ball})$$
$$v'_{girl} = -\frac{m_{ball}v'_{ball}}{m_{girl}}$$
$$v'_{girl} = -\frac{(0.60\ \text{kg})(+10\ \text{m/s})}{(60\ \text{kg})}$$
$$v'_{girl} = -0.10\ \text{m/s} \quad \text{or} \quad -10\ \text{cm/s}$$

The horizontal component of her velocity is -10 cm/s, or 10 cm/s to the left. She moves with a speed of 10 cm/s.

LESSON

When we use the momentum of two (or more) objects, any forces they exert on each other cancel. We can hide the details of their interaction this way. Therefore, conservation of momentum is usually the best way to handle a collision.

EXAMPLE

A 6000 kg truck coasts at a speed of 4.0 m/s on a horizontal road. As it passes underneath you drop a 1500 kg crate onto the truck. What is the speed of the truck after the crate is moving with the truck?

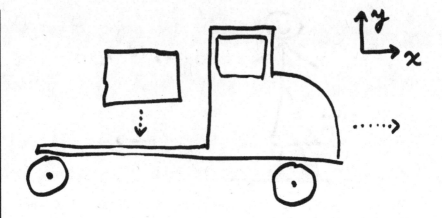

When the crate first lands on the truck, the truck is moving but the crate is not (horizontally). The truck bed will slide underneath the crate. Friction will push the crate forward (kinetic friction because the surfaces are sliding), while pushing the truck backward. The crate will speed up and the truck will slow down until their velocities match, when they won't be sliding anymore.

We don't know the force of friction, so using forces to find the accelerations won't work. As the crate slides on the truck some of the mechanical energy is turned into heat, and since we don't know how much we won't get anywhere using conservation of energy. Let's try conservation of momentum.

$$P + \Delta P = P'$$

Because momentum is a vector, we divide the momentum into x and y components and do them separately.

$$P_x + \Delta P_x = P'_x \quad \text{and} \quad P_y + \Delta P_y = P'_y$$

Since we want to find the x velocity of the truck afterward, we concentrate on the x equation.

$$(m_T v_{x,T} + m_C v_{x,C}) + (F_{x,T \text{ on } C}t - F_{x,C \text{ on } T}t) = (m_T v'_{x,T} + m_C v'_{x,C})$$

As before, the forces of the truck on the crate and the crate on the truck are equal but opposite, so that they add to zero. The crate is not moving in the x direction before landing on the truck, so $v_{x,C} = 0$.

$$m_T v_{x,T} = \left(m_T v'_{x,T} + m_C v'_{x,C}\right)$$

The truck and the crate have the same velocity afterward, so $v'_{x,T} = v'_{x,C}$.

$$m_T v_{x,T} = (m_T + m_C) v'_{x,T}$$

$$v'_{x,T} = \frac{m_T v_{x,T}}{(m_T + m_C)}$$

$$v'_{x,T} = \frac{(6000 \text{ kg})(4.0 \text{ m/s})}{(6000 \text{ kg}) + (1500 \text{ kg})}$$
$$v'_{x,T} = 3.2 \text{ m/s}$$

We don't even need to do the y part of the problem to get our answer. We'll do it anyway to see what we might have learned.

$$(m_T v_{y,T} + m_C v_{y,C}) + (F_{y,T \text{ on } C} t - F_{y,C \text{ on } T} t + F_{y,\text{road on } T} t - F_{y,\text{gravity}} t)$$
$$= (m_T v'_{y,T} + m_C v'_{y,C})$$

The forces of the truck and crate on each other cancel, $v_{y,T} = 0$, and $v'_{y,T} = v'_{y,C} = 0$.

$$(m_C v_{y,C}) + (F_{y,\text{road on } T} t - F_{y,\text{gravity}} t) = 0$$

During the landing of the crate on the truck, the road must have pushed upward on the truck harder than gravity pushed down on the truck and crate.

LESSON

Because momentum has direction, splitting the momentum into components lets us ignore forces perpendicular to our direction.

6.3 IMPULSE AND FORCE

By looking at the momentum of a pair of objects, we were able to eliminate the forces that they applied to each other. If we want that force, or if that force is the information we know, then we don't want it to cancel. Then we have to look at the momentum of just one of the two objects.

EXAMPLE

A 60 kg girl on roller blades throws a 0.60 kg ball, giving it a speed of 10 m/s. If it takes her 0.14 s to throw the ball, what is the average force that she exerts on the ball?

In a similar problem in the last section, we used conservation of momentum on her and the ball. This caused the force she applied to the ball to cancel when added to the force the ball applied to her. Since we want this force, that technique won't work. Instead, we do conservation of momentum on just the ball.

$$P + \Delta P = P'$$

$$P_x + \Delta P_x = P'_x$$

$$m_{ball}v_{ball} + F_{girl\ on\ ball}t = m_{ball}v'_{ball}$$

$$m_{ball}(0) + F_{girl\ on\ ball}t = m_{ball}v'_{ball}$$

$$F_{girl\ on\ ball} = m_{ball}v'_{ball}/t$$

$$F_{girl\ on\ ball} = (0.60\ kg)(10\ m/s)/(0.14\ s)$$

$$F_{girl\ on\ ball} = 43\ N$$

The force is positive, indicating that the force is to the right in the figure.

EXAMPLE

A 300 g rubber ball is dropped from an open window. It hits the pavement below with a speed of 30 m/s and bounces back with a speed of 20 m/s. If the collision lasts 0.025 s, what was the average force of the pavement on the ball?

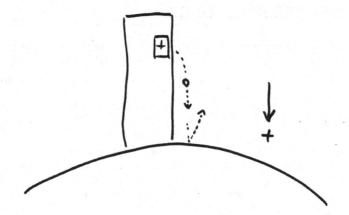

We want the force, but we don't know that the force is constant or that the acceleration is constant. (It won't be, because the more the bottom of the ball is smushed in, the greater the force.) The nonconstant acceleration tells us to try a conservation law. The kinetic energy of the ball changes because the ground does negative work on the ball. That means the force that the ground exerts on the ball as the ball goes down is more than the force it exerts as the ball goes up. Since we don't know either of these forces, and we don't know the distance involved, conservation of energy doesn't look promising, so we'll try conservation of momentum. Similar to the last example, we want the force that the two objects (ball and sidewalk) apply to each other, so we do conservation of momentum on just the ball.

$$P + \Delta P = P'$$
$$P_y + \Delta P_y = P'_y$$
$$m_{ball}v_{ball} + F_{pavement \ on \ ball}t = m_{ball}v'_{ball}$$
$$F_{pavement \ on \ ball} = m_{ball} \ (v'_{ball} - v_{ball})/t$$
$$F_{pavement \ on \ ball} = (0.300 \ \text{kg}) \ [(-20 \ \text{m/s}) - (+30 \ \text{m/s})] \ /(0.025 \ \text{s})$$
$$F_{pavement \ on \ ball} = -600 \ \text{N}$$

The force is negative, indicating that the force is upward in the figure.

LESSON

If you care about the details of the interaction, use the momentum of only one object.

EXAMPLE

A 300 g rubber ball is dropped from an open window. It hits the pavement below with a speed of 30 m/s and bounces back with a speed of 20 m/s. How fast does Earth bounce back from the collision?

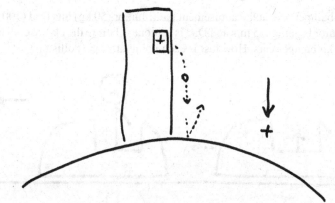

Again we use conservation of momentum, but this time on Earth.

$$P + \Delta P = P'$$
$$m_{Earth}v_{Earth} + F_{ball \ on \ Earth}t = m_{Earth}v'_{Earth}$$
$$m_{Earth}(0) + (600 \ \text{N})(0.025 \ \text{s}) = (6 \times 10^{24} \ \text{kg})v'_{Earth}$$

The 600 newtons is taken from the last example—as Earth exerts a 600 N force upward on the ball, the ball exerts a 600 N force downward on the Earth. If the collision had lasted a different length of time, the force would also have been different but the impulse, or change in momentum, would have been the same, since the change in momentum of the ball is the same. We could also have done this problem using the momentum of the two objects, as in the last section, and gotten the same result.

$$v'_{Earth} = (600 \text{ N})(0.025 \text{ s})/(6 \times 10^{24} \text{ kg})$$
$$v'_{Earth} = 2.5 \times 10^{-24} \text{ m/s}$$

This is about 0.1 micrometer per life of the universe, much too small to measure. We could also have done this problem by using the combined momentum of the ball and Earth.

6.4 ELASTIC COLLISIONS

Momentum is always conserved and energy is always conserved. Sometimes mechanical energy is turned into other forms of energy, like heat. When this happens, conservation of energy, while true, typically can't be used to solve the problem. This has been the case in every example in this chapter, so far.

Sometimes in a collision the mechanical energy is conserved. Since the time of the collision is too short for the potential energy to change, this means that the kinetic energy is the same after the collision as before, or "conserved" (not a real conservation law). We call these collisions **elastic collisions**. Some examples of such collisions include billiard balls or bumper cars striking one another.

EXAMPLE

Playing at bumper cars at the amusement park, Junior (50 kg) hits Dad (100 kg) from behind. Junior is going 3.5 m/s and Dad is stationary before the elastic collision. Both are in 100 kg bumper cars. How fast is each going after the collision?

Since this is a collision, we try conservation of momentum.

$$P + \Delta P = P'$$
$$P_x + \Delta P_x = P'_x$$
$$m_J v_J + m_D v_D + (0)t = m_J v'_J + m_D v'_D$$

Since we are only looking at horizontal momentum, we can ignore gravity and the normal force from the floor. The forces that Junior's car and Dad's car put on each other cancel when we add the impulse components.

We can't solve this, since there are two unknowns in the equation (v'_J and v'_D). We need another piece of information. The additional piece is that this is an elastic collision, so kinetic energy is the same before and after.

$$\frac{1}{2}m_J v_J^2 + \frac{1}{2}m_D v_D^2 = \frac{1}{2}m_J(v'_J)^2 + \frac{1}{2}m_D(v'_D)^2$$

We now have two equations and two unknowns, so we can solve them. I won't show all of the math, but the result is

$$v'_J = -0.5 \text{ m/s} \quad \text{and} \quad v'_D = +3.0 \text{ m/s}$$

Dad is pushed forward while Junior bounces back just a little.

Because elastic collisions are somewhat common, and because of the unpleasantness of solving simultaneous equations (especially with squares in them), physicists do something they don't often do—work out the result and reuse it the next time. If object 1 collides elastically with object 2, then their velocities afterward will be

$$v'_1 = \frac{m_1 - m_2}{m_1 + m_2}v_1 + \frac{2m_2}{m_1 + m_2}v_2$$

$$v'_2 = \frac{m_2 - m_1}{m_1 + m_2}v_2 + \frac{2m_1}{m_1 + m_2}v_1$$

or, if object 2 is stationary before the collision,

$$v'_1 = \frac{m_1 - m_2}{m_1 + m_2}v_1 \quad \text{and} \quad v'_2 = \frac{2m_1}{m_1 + m_2}v_1$$

These equations can be used to find the results in the example above. To use them, the positive direction for each object must be the same.

How do you tell if a collision is elastic?

- If the problem says that the collision is elastic, then the collision is elastic (really).
- If the two objects stick together, then the collision is not elastic.
- If one object passes through the other, then the collision is not elastic.
- If you can solve the problem using only conservation of momentum, then there is no need to worry whether the collision is elastic.
- After that, use your experience with the objects in question: are they likely to bounce off of one another (elastic) or is some of the energy likely to be turned into heat (not elastic)?

EXAMPLE

A pool (billiard) ball moving 2.4 m/s strikes another pool ball head-on. What is the velocity of each pool ball after the collision?

Pool balls typically collide elastically. Since they all have the same mass,

$$v_1' = \frac{m_1 - m_2}{m_1 + m_2}v_1 = \frac{m - m}{m + m}v_1 = 0$$

$$v_2' = \frac{2m_1}{m_1 + m_2}v_1 = \frac{2m}{m + m}v_1 = v_1 = 2.4 \text{ m/s}$$

Whenever two objects of identical mass collide elastically, they switch velocities. If they don't hit head-on but a little to the side, then we need to do a two-dimensional problem with the x and y momentum separately.

EXAMPLE

A rubber ball bounces elastically off of a brick wall. What is the velocity of each object after the collision?

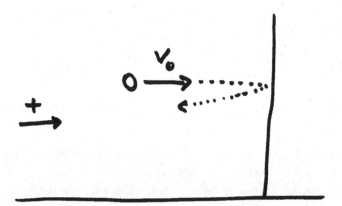

Because the collision is elastic (and one-dimensional), we use the formulas above.

$$v_1' = \frac{m_1 - m_2}{m_1 + m_2}v_1 \quad \text{and} \quad v_2' = \frac{2m_1}{m_1 + m_2}v_1$$

The mass of the rubber ball is so much smaller than the mass of the brick wall that we can use zero as the mass of the rubber ball m_1.

$$v_1' = \frac{-m_2}{m_2}v_1 \quad \text{and} \quad v_2' = \frac{0}{m_2}v_1$$

$$v_1' = -v_1 \quad \text{and} \quad v_2' = 0$$

The ball bounces back with the same speed while the brick wall doesn't move. How then is momentum conserved? The brick wall does move a little, and our answer is an approximate result. But like Earth and the rubber ball above, the speed of Earth is too small to measure.

EXAMPLE

A 46 g golf ball is hit (elastically) by a 500 g golf club moving at 40 m/s. What is the velocity of the golf ball after being struck?

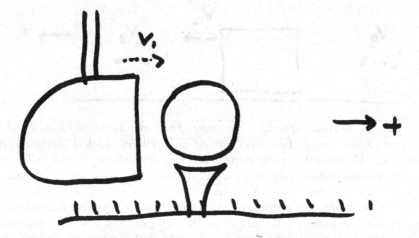

Again the collision is elastic (and one-dimensional), so we use the formulas above.

$$v_1' = \frac{m_1 - m_2}{m_1 + m_2}v_1 \quad \text{and} \quad v_2' = \frac{2m_1}{m_1 + m_2}v_1$$

$$v_2' = \frac{2(500 \text{ g})}{(46 \text{ g}) + (500 \text{ g})}(40 \text{ m/s})$$

$$v_2' = 73 \text{ m/s}$$

The ball moves faster than the clubhead was moving!

The collision of a clubhead and a golf ball is not completely elastic. Golfers use the coefficient of restitution (COR) to describe how elastic the collision is.

6.5 USING CONSERVATION OF MOMENTUM

Conservation of momentum hides the interaction between two objects, and works well for collision processes.

EXAMPLE

A 4.0 g bullet moving at 460 m/s hits a 410 g block of wood. The bullet comes out the other side at a speed of 214 m/s. The wood block slides along the horizontal surface with a coefficient of friction of 0.42. How far does the block slide?

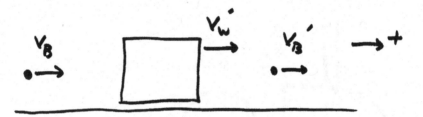

Is this an elastic collision? We might expect that the wood block would be damaged by the passing of the bullet, turning some of the mechanical (kinetic) energy into heat. Also, the bullet passing through the block of wood is a telltale sign. Because it is an inelastic collision, we can't use conservation of energy to solve the problem.

We can try conservation of momentum. If we start with the moving bullet and end with the stopped block, then we need to know how fast the planet is moving after the wood block pushes it (through friction). What we can find, using conservation of momentum, is the speed of the wood block just after the bullet passes through it. Then we could use forces and constant acceleration, or conservation of energy, to figure out how far the block slides.

We use conservation of momentum to find the speed of the wood block.

$$P + \Delta P = P'$$

$$(m_B v_B + m_W v_W) + (F_{B \text{ on } W} - F_{W \text{ on } B})t = (m_B v_B' + m_W v_W')$$

$$[m_B v_B + m_W(0)] + (0)t = (m_B v_B' + m_W v_W')$$

$$m_W v_W' = m_B v_B - m_B v_B'$$

$$v_W' = \frac{m_B(v_B - v_B')}{m_W}$$

$$v'_W = \frac{(4.0 \text{ g})}{(410 \text{ g})}[(460 \text{ m/s}) - (214 \text{ m/s})]$$

$$v'_W = 2.4 \text{ m/s}$$

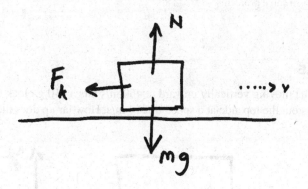

How far does the block slide? We can use forces:

$$F = ma$$
$$-F_k = ma_x$$
$$F_k = \mu N$$
$$+N - mg = ma_y = 0$$
$$N = mg$$
$$F_k = \mu mg$$
$$-\mu mg = ma_x$$
$$a_x = -\mu g$$
$$v^2 - v_0^2 = 2a(\Delta x)$$
$$(0)^2 - (2.4 \text{ m/s})^2 = 2[-(0.42)(9.8 \text{ m/s}^2)](d)$$
$$d = 0.70 \text{ m} \quad \text{or} \quad 70 \text{ cm}$$

Or we can use conservation of energy:

$$KE_{\text{before}} + PE_{\text{before}} + W = KE_{\text{after}} + PE_{\text{after}}$$
$$\frac{1}{2}m(v_0)^2 + mgh + (-\mu Nd) = \frac{1}{2}m(0)^2 + mgh$$
$$\frac{1}{2}m(v_0)^2 - \mu mgd = 0$$
$$d = \frac{(v_0)^2}{2\mu g}$$
$$d = \frac{(2.4 \text{ m/s})^2}{2(0.42)(9.8 \text{ m/s}^2)}$$
$$d = 0.70 \text{ m}$$

Conservation of momentum is a tool, a technique that can be used along with other techniques to solve problems.

EXAMPLE

A 4.0 g bullet moving vertically upward at 460 m/s hits a 410 g block of wood. The bullet comes out the top side at a speed of 214 m/s. How far up does the wood block go?

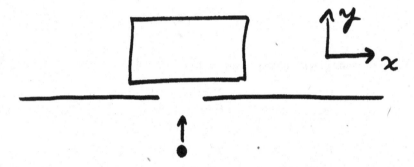

As in the previous example, we can use momentum to find the speed of the block just after the bullet passes through, then use another technique to figure out how far it goes.

$$P + \Delta P = P'$$

$$(m_B v_B + m_W v_W) + [F_{B \text{ on } W} - F_{W \text{ on } B} - (m_B + m_W)g]t = (m_B v'_B + m_W v'_W)$$

The forces of the bullet and the wood on each other cancel out, and v_W is zero.

$$m_B v_B + [-(m_B + m_W)g]t = (m_B v'_B + m_W v'_W)$$

The time of the collision is very short,

$$t \approx \text{(height of block)}/v \approx (1 \text{ m})/(330 \text{ m/s}) \approx 0.3 \text{ s}$$

A 410 g block is unlikely to be more than 1 m high. The fractional change in the momentum due to gravity during the collision is

$$\frac{(mg)(t)}{m_B v_B} \approx \frac{(4 \text{ N})(3 \times 10^{-4} \text{ s})}{(4.0 \times 10^{-3} \text{ kg})(460 \text{ m/s})} \approx 6.5 \times 10^{-4} \quad \text{or} \quad 0.07\%$$

The impulse (change in the momentum) from gravity during the collision is so small that we'll ignore it.

$$m_B v_B = m_B v'_B + m_W v'_W$$

$$v'_W = m_B(v_B - v'_B)/m_W$$

$$v'_W = (4.0 \text{ g})[(460 \text{ m/s}) - (214 \text{ m/s})]/(410 \text{ g})$$
$$v'_W = 2.4 \text{ m/s}$$

How high does the block go? We can use constant acceleration:

$$v^2 - v_0^2 = 2a \, \Delta x$$
$$(0)^2 - (2.4 \text{ m/s})^2 = 2(-9.8 \text{ m/s}^2)h$$
$$h = \frac{(2.4 \text{ m/s})^2}{2(9.8 \text{ m/s}^2)}$$
$$h = 0.29 \text{ m} \quad \text{or} \quad 29 \text{ cm}$$

We could also use conservation of energy:

$$KE_{before} + PE_{before} + W = KE_{after} + PE_{after}$$
$$\frac{1}{2}m(v_0)^2 + mg(0) + (0) = \frac{1}{2}m(0)^2 + mgh$$
$$h = \frac{(v_0)^2}{2g}$$
$$h = \frac{(2.4 \text{ m/s})^2}{2(9.8 \text{ m/s}^2)}$$
$$h = 0.29 \text{ m}$$

LESSON

Collision times are usually short enough that we ignore other forces during the collision.

EXAMPLE

A 0.85 kg ball on the end of a 35 cm long string falls and hits a 0.56 kg block of putty. The block sticks to the ball. What is the speed of the putty after it is hit by the ball?

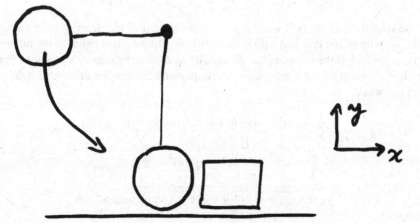

We have used conservation of momentum to handle many collisions in this chapter, so we'll probably want to use conservation of momentum again.

$$P + \Delta P = P'$$

$$(m_B v_B + m_P v_P) + (F_{\text{ball on putty}} - F_{\text{putty on ball}})t = (m_B v'_B + m_P v'_P)$$

Here we ignore any horizontal friction between the putty and the floor during the collision (the collision time is short so the impulse from friction *during* the collision is very small, like gravity in the last example). The forces of the ball and the putty on each other cancel (that's why we use momentum for collisions), $v_P = 0$, and $v'_B = v'_P$.

$$m_B v_B = (m_B + m_P) v'_P$$

We know the masses, so to find v'_P we need to know v_B.

Can we use conservation of momentum to find v_B, the (horizontal) speed of the ball just before the collision? The forces acting on the ball are gravity and the tension in the string. Gravity is always vertical and we're using the horizontal component of the momentum, so we can ignore gravity. The tension force changes not only in direction but also in magnitude as the ball falls. If we knew what this tension force was at all times during this process, we could integrate it (ugh!) to find the horizontal impulse from the string to the ball. This would be ugly. Alternatively, the other end of the string is attached to Earth, so if we knew how fast Earth was moving to the left after the ball fell, we could use momentum to find the speed of the ball just before the collision. But that speed is far too small to measure. Conservation of momentum doesn't work for this—we need to try something else.

The acceleration of the ball isn't constant, and we don't know (or care about) either the acceleration or the time involved. This sounds like the characteristics of problem that can be solved using conservation of energy.

$$KE_{\text{before}} + PE_{\text{before}} + W = KE_{\text{after}} + PE_{\text{after}}$$

$$\frac{1}{2} m_B (0)^2 + m_B g h + W = \frac{1}{2} m_B (v)^2 + m_B g(0)$$

The velocity v of the ball when it gets to the bottom of the quarter-circle swing is the velocity of the ball just before the collision. The height is equal to the radius of the arc, which is the length of the string. The work done by gravity is included in the potential energy, and the tension is always perpendicular to the motion so it doesn't do any work.

$$\frac{1}{2} m_B (0)^2 + m_B g R + 0 = \frac{1}{2} m_B (v_B)^2 + m_B g(0)$$

$$g R = \frac{1}{2} (v_B)^2$$

$$v_B = \sqrt{2gR} = \sqrt{2(9.8 \text{ m/s}^2)(0.35 \text{ m})} = 2.62 \text{ m/s}$$

Now we can find the speed of the putty (and ball) after the collision.

$$m_B v_B = (m_B + m_P) v_P'$$

$$v_P' = \frac{m_B v_B}{m_B + m_P}$$

$$v_P' = \frac{0.85 \text{ kg}}{0.85 \text{ kg} + 0.56 \text{ kg}} \, (2.62 \text{ m/s})$$

$$v_P' = 1.58 \text{ m/s}$$

EXAMPLE

A 0.85 kg, 35 cm long uniform rod falls from vertical to horizontal, pivoting about its end, then strikes a 0.56 kg block of putty. The block sticks to the end of the rod. What is the speed of the putty after it is hit by the rod?

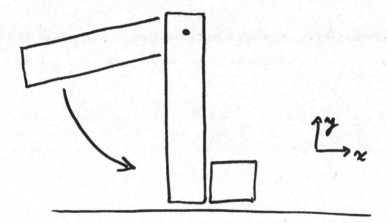

This problem looks and sounds remarkably similar to the last one, so we'll try to solve it using the techniques that worked so well. We use conservation of energy to find the speed of the rod as the objects collide, then use conservation of momentum to find the speed of the putty after the collision.

The difficulty with the first step is that not all of the rod will be moving with the same speed. We can't even use the speed of the center of mass, because kinetic energy is proportional to v^2 and isn't linear. We'd like to use conservation of energy for this, but we need to know how to find the kinetic energy of something that's rotating (see next chapter).

Even if we could do the first step, we'd have another difficulty. During the collision, the pivot at the top of the rod will exert a force to keep the top of the rod in place. Earlier we ignored other forces like this because the collision time was small. That doesn't apply here because the smaller the collision time, the larger the force from the pivot. The product of the two, which is the impulse on the rod, will be a constant. Unless we know that impulse, we won't be able to solve the second step using conservation of momentum. We will be able to solve this using conservation of angular momentum (next chapter).

LESSON

When there is a constraint (like a pivot or hinge), the force it exerts during a collision is too large to ignore. Conservation of momentum doesn't work well for such problems.

CHAPTER SUMMARY

- Momentum is conserved, so

$$\overbrace{P}^{m\boldsymbol{v}} + \underbrace{\Delta P}_{F_t} = P'$$

- Momentum is a vector, so we can treat each component separately, or not at all.

- Conservation of momentum "hides" the forces between objects, so it works well for collisions.

CHAPTER 7

ROTATION

Sometimes things move, but sometimes things rotate (and talented objects can do both at the same time). Everything we've done so far also applies for rotation, using the same equations, but with different variables.

x	$\rightarrow$	θ (Greek theta)	angular displacement
v	$\rightarrow$	ω (Greek omega)	angular velocity
a	$\rightarrow$	α (Greek alpha)	angular acceleration
m	$\rightarrow$	I	angular mass = moment of inertia
F	$\rightarrow$	τ (Greek tau)	angular force = torque
P	$\rightarrow$	L	angular momentum

Every equation we've had will also have an angular equivalent. For example, force equals mass times acceleration ($F = ma$), so angular force equals angular mass times angular acceleration ($\tau = I\alpha$).

7.1 CONSTANT ANGULAR ACCELERATION

An object can have a velocity, which is the change in its position. It can have an acceleration, which is how fast the velocity is changing. It can have an acceleration even if the velocity is zero: throw something (other than this book) straight up, and at the instant that it is at its peak height the velocity is zero but changing, so that there is an acceleration. We've done all this before.

It's all true for rotation as well. An object can have an angular velocity, which is the change in its angular position or angular displacement, or how fast it is spinning. It can have an angular acceleration, which is how fast the angular velocity is changing— is the rotation speeding up or slowing down? It can have an angular acceleration even if the angular velocity is zero. Roll a basketball or other round object up a ramp, and at the instant that it is at its peak the angular velocity is zero but changing, it was rotating one way, is not rotating now, and is about to rotate the other way. There is an angular acceleration.

If the acceleration were constant then we could solve a number of problems with a few short equations. We could solve problems where the acceleration wasn't constant, but that was harder. All of the equations we used for constant acceleration work for constant angular acceleration as well—just replace the variable with the angular equivalent.

$$v - v_0 = at \qquad \rightarrow \qquad \omega - \omega_0 = \alpha t$$
$$\Delta x = v_0 t + \tfrac{1}{2} a t^2 \qquad \rightarrow \qquad \Delta\theta = \omega_0 t + \tfrac{1}{2} \alpha t^2$$
$$\Delta x = \tfrac{1}{2}(v_0 + v)t \qquad \rightarrow \qquad \Delta\theta = \tfrac{1}{2}(\omega_0 + \omega)t$$
$$v^2 - v_0^2 = 2a\,\Delta x \qquad \rightarrow \qquad \omega^2 - \omega_0^2 = 2\alpha\,\Delta\theta$$

The angles are measured in radians. What is a radian? I'll tell you later. For now, 2π of them is one circle, rotation, or revolution.

EXAMPLE

A 0.82 m diameter wheel rotates about its center with an angular velocity of 32 rad/s. What angular acceleration is needed to stop it in 5.0 s?

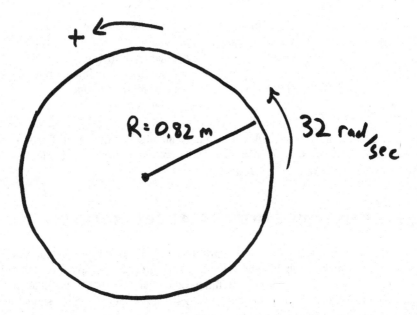

We attack this the same way we did constant acceleration:

Δx	$\rightarrow$	$\Delta\theta$	$=$	
v_0	$\rightarrow$	ω_0	$=$	32 rad/s
v	$\rightarrow$	ω	$=$	0
a	$\rightarrow$	α	$=$	?
t	$\rightarrow$	t	$=$	5.0 s

We don't care about the angular displacement, so we choose

$$\omega - \omega_0 = \alpha t$$
$$\alpha = (\omega - \omega_0)/t$$
$$\alpha = [(0) - (32 \text{ rad/s})]/(5.0 \text{ s})$$
$$\alpha = -6.4 \text{ rad/s}^2$$

The angular acceleration is negative. Because the rotation of the wheel is slowing, the angular velocity ω and the angular acceleration α must be in opposite directions, one positive and the other negative.

EXAMPLE

A turntable starting from rest rotates 5 times while accelerating to 48 rpm (rotations per minute). What is the angular acceleration of the turntable?

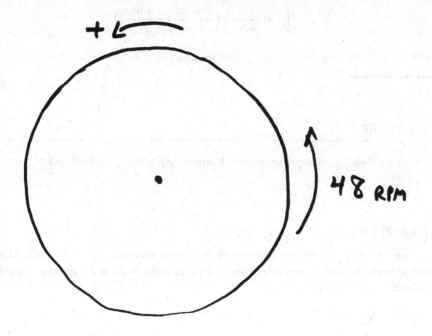

We use our constant acceleration technique:

Δx	$\rightarrow$	$\Delta\theta$	$=$	5 rotations
v_0	$\rightarrow$	ω_0	$=$	0
v	$\rightarrow$	ω	$=$	48 rot/min
a	$\rightarrow$	α	$=$	?
t	$\rightarrow$	t	$=$	

We don't care about the time, so we choose

$$\omega^2 - \omega_0^2 = 2\alpha\,\Delta\theta$$

$$\alpha = \frac{\omega^2 - \omega_0^2}{2\,\Delta\theta}$$

$$\alpha = \frac{(48 \text{ rot/min})^2 - (0)^2}{2(5 \text{ rot})}$$

$$\alpha = 230 \text{ rot/min}^2$$

Every minute the angular velocity increases (becomes more positive) by 230 rotations per minute.

Usually the angular acceleration is given in radians/second^2. Sometimes it is necessary to use radians—I'll explain in the next section. We'll convert (for practice, mostly).

$$\alpha = \left(230\frac{\text{rot}}{\text{min}^2}\right)\left(\frac{2\pi \text{ rad}}{1 \text{ rot}}\right)\left(\frac{1 \text{ min}}{60 \text{ s}}\right)^2$$

$$\alpha = 0.40 \text{ rad/s}^2$$

LESSON

It is not always necessary to use radians. It is necessary, as always, to keep track of your units.

EXAMPLE

A wheel beginning at 51 rpm rotates with an angular acceleration of 3.7 rad/s^2. After 4.5 seconds, it slows to rest over a time of 16 seconds. What is the average angular velocity of the wheel?

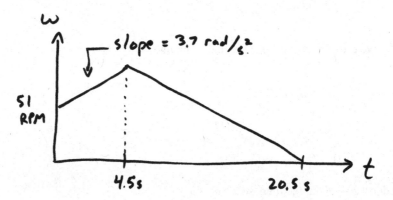

This problem has two different angular accelerations in it, one positive and one negative. Because the accelerations are not the same, but each step has a constant acceleration (piecewise constant), we can solve each part using constant angular acceleration techniques.

Before we do so, consider what we're trying to find. We want the average angular velocity. If we could find the average angular velocity for each part, would the average angular velocity be the average of the two? Not necessarily, because the time for each "half" of the problem is not the same. Instead, the average angular velocity is

$$\omega_{ave} = \frac{\Delta\theta}{t} = \frac{\Delta\theta_1 + \Delta\theta_2}{t_1 + t_2}$$

We need to find how far the wheel rotates for each part of the problem, add to find the total angular displacement (or distance, since the wheel is always rotating the same direction), then divide by the total time.

For the first part:

Δx	$\rightarrow$	$\Delta\theta$	$=$	?
v_0	$\rightarrow$	ω_0	$=$	51 rpm
v	$\rightarrow$	ω	$=$	
a	$\rightarrow$	α	$=$	3.7 rad/s^2
t	$\rightarrow$	t	$=$	4.5 s

We don't care about the final angular velocity, so we use

$$\Delta\theta = \omega_0 t + \frac{1}{2}\alpha t^2$$

$$\Delta\theta_1 = \left[(51 \text{ rot/min}) \left(\frac{2\pi \text{ rad}}{1 \text{ rot}} \right) \left(\frac{1 \text{ min}}{60 \text{ s}} \right) \right] (4.5 \text{ s}) + \frac{1}{2}(3.7 \text{ rad/s}^2)(4.5 \text{ s})^2$$

$$\Delta\theta_1 = (5.34 \text{ rad/s}) (4.5 \text{ s}) + \frac{1}{2}(3.7 \text{ rad/s}^2)(4.5 \text{ s})^2$$

$$\Delta\theta_1 = 61.5 \text{ rad}$$

For the second part:

Δx	$\rightarrow$	$\Delta\theta$	$=$	?
v_0	$\rightarrow$	ω_0	$=$	ω_1
v	$\rightarrow$	ω	$=$	0
a	$\rightarrow$	α	$=$	
t	$\rightarrow$	t	$=$	16 s

We need to find the angular velocity at the end of the first step and use that as the initial angular velocity for the second step. ·

$$\omega - \omega_0 = \alpha t$$

$$\omega = \omega_0 + \alpha t$$

$$\omega = (5.34 \text{ rad/s}) + (3.7 \text{ rad/s}^2)(4.5 \text{ s})$$

$$\omega = 22.0 \text{ rad/s}$$

We don't care about the angular acceleration during the second step, so we use

$$\Delta\theta_2 = \frac{1}{2}(\omega_0 + \omega)t$$

$$\Delta\theta_2 = \frac{(22.0 \text{ rad/s}) + (0)}{2}(16 \text{ s})$$

$$\Delta\theta_2 = 176 \text{ rad}$$

The average angular velocity is

$$\omega_{\text{ave}} = \frac{\Delta\theta_1 + \Delta\theta_2}{t_1 + t_2}$$

$$\omega_{\text{ave}} = \frac{(61.5 \text{ rad}) + (176 \text{ rad})}{(4.5 \text{ s}) + (16 \text{ s})}$$

$$\omega_{\text{ave}} = 11.6 \text{ rad/s}$$

LESSON

Solve constant angular acceleration problems the same way you solve constant acceleration problems: write down the five symbols and identify three of them.

7.2 CONNECTING ANGULAR AND LINEAR MOTION

Sometimes we need to connect the angular motion with the linear motion. Imagine, for example, that we put a small drop of paint on the wheel before it rotates. We might then ask, "How fast is the drop of paint moving?" Note that the velocity of the paint is constantly changing, because its direction is changing, but if the angular velocity is constant then the speed of the paint drop is also constant.

The further out from the pivot point (or axle) the paint is, the further it must move in each rotation, and the faster we expect it to move. Since the distance covered is $2\pi R$ for each rotation, the speed is

$$v = 2\pi R \times (\text{rotations per second})$$

By choosing the unit for angles to be 2π radians per rotation, this becomes

$$\Delta x = R\,\Delta\theta$$
$$v = R\omega$$
$$a = R\alpha$$

If we had chosen some other unit for measuring angles, then the above formulas would need to include an extra constant; by using radians we get beautiful, simple formulas. This means that we *must* use radians whenever we use one of the formulas above. Sometimes we use one of them without knowing it, so the general rule is that we use radians whenever there is a length in the problem—any length.

EXAMPLE

In the previous example ("A wheel beginning at 51 rpm ..."), what is the average speed of a point on the outside of the wheel? The radius of the wheel is 62 cm.

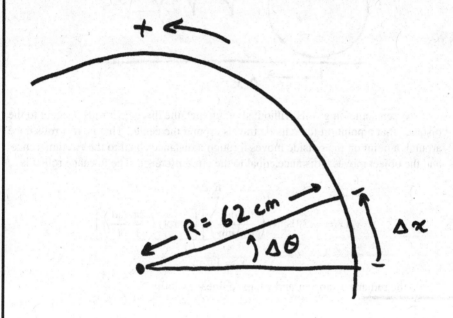

The speed of a point on the outside is

$$v = R\omega$$

so the average speed is the distance to the pivot point times the angular speed.

$$v = \left[(62 \text{ cm}) \left(\frac{1 \text{ m}}{100 \text{ cm}} \right) \right] (11.6 \text{ rad/s})$$

$$v = 7.2 \text{ m/s}$$

What happened to the "radians"? Shouldn't (m) times (rad/s) be (m rad/s)? Radians aren't really a unit, so we can add them when we want and remove them when we don't. To say an angle is "0.14" is the same as to say "0.14 radians." We can't do this with rotations, degrees, or any other unit.

EXAMPLE

A 6.0 cm diameter tube rolls without slipping 9 times across a horizontal surface. How far does it roll?

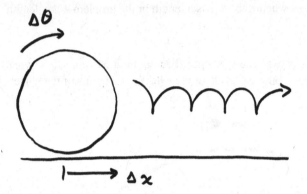

When something rolls without slipping then the distance it rolls is equal to the distance that a point on the outside travels around the center. That is, if it rolls once around, a point on the outside moves through a distance equal to the circumference, and the object moves a distance equal to the circumference. The distance rolled is

$$\Delta x = R \, \Delta\theta$$

$$\Delta x = \left[(3.0 \text{ cm}) \left(\frac{1 \text{ m}}{100 \text{ cm}}\right)\right]\left[(9 \text{ rot}) \left(\frac{2\pi \text{ rad}}{1 \text{ rot}}\right)\right]$$

$$\Delta x = 1.7 \text{ m}$$

Again the radians disappear and we get a linear quantity.

LESSON

Use radians whenever there is a length in the problem, any length.

EXAMPLE

A 42 cm diameter wheel is rotating with an angular speed of 120 rpm. It is slowing with an angular deceleration of 52 rad/s². What is the magnitude of the (linear) acceleration of a point on the edge of the wheel?

We have two equations for the acceleration that we might use.

$$a = \frac{v^2}{R} \quad \text{and} \quad a = R\alpha$$

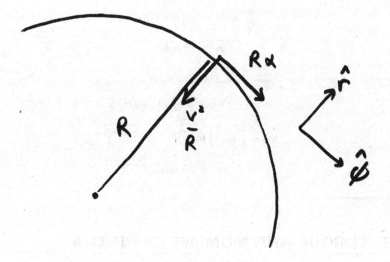

The first one is the acceleration in toward the middle of something going in a circle. Because the direction of the velocity of our "paint drop" is changing, it must be accelerating, and this acceleration is toward the middle of the circular path and is v^2/R.

The second is the acceleration that causes our paint drop to move faster. As the wheel spins faster and faster, the paint drop moves faster and faster. This acceleration, in the same direction as the motion, causes the paint drop to speed up and points in the same direction as the velocity. This acceleration is $a = R\alpha$.

The acceleration is

$$a = a_R\hat{r} + a_T\hat{\phi} \quad \text{with} \quad a_R = -\frac{v^2}{R} \text{ and } a_T = R\alpha$$

where $\hat{r}$ is the unit vector pointing out from the center and $\hat{\phi}$ is the unit vector pointing tangent to the circle, perpendicular to $\hat{r}$.

$$a_R = -\frac{v^2}{R}$$

$$a_R = -\frac{(R\omega)^2}{R}$$

$$a_R = -R\omega^2$$

$$a_R = -\left[(21 \text{ cm})\left(\frac{1 \text{ m}}{100 \text{ cm}}\right)\right]\left[(120 \text{ rot/min})\left(\frac{2\pi \text{ rad}}{1 \text{ rot}}\right)\left(\frac{1 \text{ min}}{60 \text{ s}}\right)\right]^2$$

$$a_R = -33 \text{ m/s}^2$$

$$a_T = R\alpha$$

$$a_T = (0.21 \text{ m}) (52 \text{ rad/s}^2)$$

$$a_T = 11 \text{ m/s}^2$$

$$a = a_R \hat{r} + a_T \hat{\phi}$$

$$a = (-33 \text{ m/s}^2)\hat{r} + (11 \text{ m/s}^2)\hat{\phi}$$

$$|a| = \sqrt{(-33 \text{ m/s}^2)^2 + (11 \text{ m/s}^2)^2}$$

$$|a| = 35 \text{ m/s}^2$$

7.3 TORQUE AND MOMENT OF INERTIA

Instead of force, we use the angular equivalent of angular force, or torque. Try opening a door by pushing near the hinge rather than at the handle. You will find that the door does not open as easily or quickly, or perhaps even at all. The farther away from the pivot point the force is applied the greater the torque.

The torque a force creates is equal to the force times the moment arm times the cosine of the angle between them:

$$\tau = FR \sin \phi$$

The "moment arm" is the line from the pivot point to the spot where the force is applied. If the force is parallel or antiparallel to the moment arm, it does not cause the object to spin and the torque is zero. Some people use θ for the angle between the force and the moment arm, but we'll use ϕ (Greek phi) since θ has already been used this chapter.

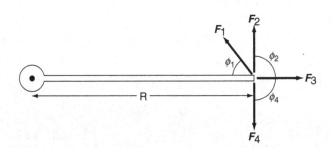

Torques can be positive or negative. Like velocities and accelerations, we get to pick the positive direction and a negative torque is one in the other direction. Once we choose the positive direction we need to stick with that direction for the whole problem. Physicists usually choose counterclockwise as the positive direction, but it's not necessary to do that. If we do, then in the figure above we would say that F_4 causes a torque in the negative direction.

EXAMPLE

A boy leaving a store pushes on the door handle with a force of 18 N. The door is 0.78 m wide and has a mass of 7.2 kg. The boy pushes perpendicular to the surface of the door. What is the torque he applies to the door?

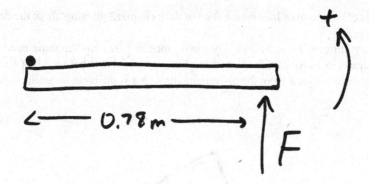

The torque is

$$\tau = FR \sin \phi$$
$$\tau = (18 \text{ N})(0.78 \text{ m}) \sin 90°$$
$$\tau = (18 \text{ N})(0.78 \text{ m})(1)$$
$$\tau = 14 \text{ N m}$$

In the energy chapter we learned that a newton times a meter is a joule. Because work and torque are both a force times a distance, they have the same units. Despite this, we would never say the torque was "14 joules" but instead say "14 newton meters," even though the two are equal. Be careful to write N m and not mN, because mN is a millinewton, or a force of one-thousandth of a newton.

In the previous example, how does the mass of the door affect the answer? The torque the boy applies is the same regardless of the mass of the door. The more massive the door, the more inertia it has and the slower it will open, but the torque is the same.

Just as Newton's second law says $F = ma$, so the same equation holds with the angular equivalents: $\tau = I\alpha$, or angular force equals angular mass times angular acceleration. The angular mass is called "moment of inertia," a combination of moment arm and inertia. The moment of inertia is equal to

$$I = \sum_i m_i r_i^2$$

We divide the object into tiny little pieces, then for each piece multiply its mass times the square of how far from the pivot point it is, then add all of those together. We need to use lots of tiny pieces so that each piece is all the same distance from the pivot point.

EXAMPLE⋆

What is the moment of inertia of a slice of uniform pizza pivoting about the center of the pizza?

By uniform we mean that any square inch of pizza has the same mass as any other square inch. We can divide the slice into pieces as shown below. All of each piece is the same distance x from the pivot point, though x is different for each piece of the slice.

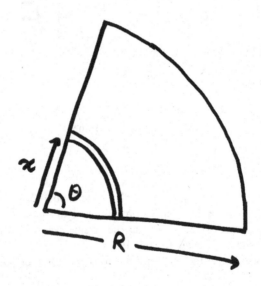

The moment of inertia is

$$I = \sum_i m_i r_i^2$$

The distance of each piece from the pivot r is x. The mass of each piece is

$$\frac{\text{area of that piece}}{\text{area of the whole slice}} \text{ (mass of the whole slice)}$$

$$I = \sum_i \left(\frac{A_i}{A}M\right)x_i^2$$

Each piece is a small (bent) rectangle with a height that increases with x. The width of each small piece is the distance between pieces dx.

$$I = \int \underbrace{(x\theta)\,dx}_{\text{area of small piece}} \frac{M}{A}x^2$$

We do this as x increases from 0 to R.

$$I = \frac{M\theta}{A}\int_0^R x^3 dx$$

$$I = \frac{M\theta}{A}\left[\frac{1}{4}x^4\right]_0^R$$

$$I = \frac{M\theta}{4A}\left(R^4 - 0^4\right)$$

$$I = \frac{M\theta R^4}{4A}$$

We substitute the area of the slice $A = (\theta/2\pi)(\pi R^2)$,

$$I = \frac{M\theta R^4}{4(\frac{1}{2}\theta R^2)}$$

$$I = \frac{MR^2}{2}$$

It doesn't matter how much of the pizza we have, so long as we know the mass of our slice.

After all of the work in the last example, you might be surprised by the simplicity of the result. This is not uncommon when finding moments of inertia. Many common shapes have simple moments of inertia, and it's much easier to look them up than to figure them out every time. See Table 7.1 for common shapes.

Sometimes, though, you might want to do something a little different. The moment of inertia of a circle (or disk) about its center is $I = \frac{1}{2}MR^2$, but what if it rotates about a point on its edge rather than its center? The **parallel axis theorem** tells us that

$$I = I_{CM} + Md^2$$

or the moment of inertia about any point on an object is equal to the moment of inertia about its center of mass plus its mass times the distance the pivot point is from the center of mass. So a disk rotated about a point on its edge would have a moment of inertia of $I = I_{CM} + Md^2 = \frac{1}{2}MR^2 + MR^2 = \frac{3}{2}MR^2$.

TABLE 7.1: Moments of Inertia for Common Objects Rotated about Their Centers

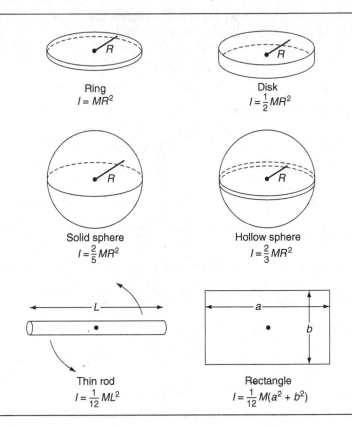

Ring
$I = MR^2$

Disk
$I = \frac{1}{2}MR^2$

Solid sphere
$I = \frac{2}{5}MR^2$

Hollow sphere
$I = \frac{2}{3}MR^2$

Thin rod
$I = \frac{1}{12}ML^2$

Rectangle
$I = \frac{1}{12}M(a^2 + b^2)$

EXAMPLE

A boy leaving a store pushes on the door handle with a force of 18 N. The door is 0.78 m wide and has a mass of 7.2 kg. The boy pushes perpendicular to the surface of the door. How long does it take for the door to open (i.e., to rotate to 90°)?

We found in an earlier example that the torque is

$$\tau = FL \sin 90° = 14\,\text{N m}$$

If we knew the moment of inertia, or angular mass, we could find the angular acceleration. From the (constant) angular acceleration we could figure out how long it takes the door to rotate 90° or $\pi/2$ radians.

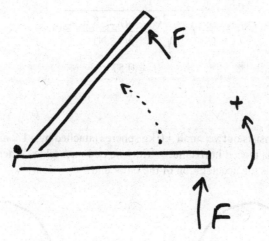

The door (as seen from above) is essentially a rod rotated about its end.

$$I = I_{CM} + Md^2$$

$$I = \frac{1}{12}ML^2 + M(L/2)^2 = \frac{1}{3}ML^2$$

The angular acceleration is

$$\tau = I\alpha$$

$$\alpha = \frac{\tau}{I} = \frac{FL\sin 90°}{\frac{1}{3}ML^2} = \frac{3F}{ML}$$

We use our constant angular acceleration technique:

Δx	$\rightarrow$	$\Delta\theta$	$= \pi/2$
v_0	$\rightarrow$	ω_0	$= 0$
v	$\rightarrow$	ω	$=$
a	$\rightarrow$	α	$= 3F/ML$
t	$\rightarrow$	t	$= ?$

We don't care about the final angular velocity, so we use

$$\Delta\theta = \omega_0 t + \frac{1}{2}\alpha t^2$$

$$\left(\frac{\pi}{2}\right) = (0)t + \frac{1}{2}\left(\frac{3F}{ML}\right)t^2$$

$$t = \sqrt{\frac{ML\pi}{3F}}$$

$$t = \sqrt{\frac{(7.2 \text{ kg})(0.78 \text{ m})\pi}{3(18 \text{ N})}}$$

$$t = 0.57 \text{ s}$$

EXAMPLE

A dumbbell consists of two small 4.0 kg spheres attached by a 1.2 m long, 2.0 kg rod. The dumbbell is placed horizontally with one end on a table. When the dumbbell is released, what is the acceleration of the other end?

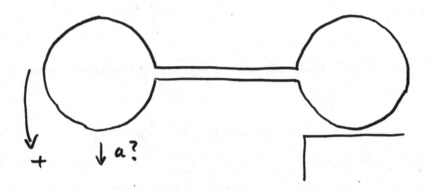

We might expect that the answer is g. However, a free body diagram of the left end shows that the rod is creating a force on the end, and this force could be vertical as well as horizontal. Since we don't know what this force is, we can't add up the forces and apply Newton's second law.

Because one end stays at a fixed location, this looks like a rotation problem. We could find the torque about the fixed end, find the moment of inertia, and calculate the angular acceleration. We then connect the acceleration with the angular acceleration using $a = R\alpha$.

The torque is

$$\tau = FR \sin \phi$$

$$\tau = (4 \text{ kg})g(1.2 \text{ m}) \sin 90° + (2 \text{ kg})g(0.6 \text{ m}) \sin 90° + (4 \text{ kg})g(0 \text{ m}) \sin(?)$$

$$\tau = 58.8 \text{ N m}$$

The moment of inertia is

$$I = I_{\text{left}} + I_{\text{rod}} + I_{\text{right}}$$

$$I = \left(M_{\text{left}}L^2\right) + \left[\frac{1}{12}M_{\text{rod}}L^2 + M_{\text{rod}}\left(\frac{L}{2}\right)^2\right] + \left[M_{\text{right}}(0)^2\right]$$

The word "small" is physics-speak for "so small that we can treat it as zero," rather than using the formula for a solid sphere, with the parallel axis theorem.

$$I = [(4 \text{ kg})L^2] + \left(\frac{1}{3}(2 \text{ kg})L^2\right)$$

$$I = \left(4 + \frac{2}{3}\right) \text{ kg} \, (1.2 \text{ m})^2$$

$$I = 6.72 \text{ kg m}^2$$

The angular acceleration is

$$\tau = I\alpha$$

$$\alpha = \frac{\tau}{I} = \frac{59 \text{ N m}}{6.7 \text{ kg m}^2} = 8.75 \text{ rad/s}^2$$

The acceleration of the end is

$$a = R\alpha = (1.2 \text{ m})(8.75 \text{ rad/s}^2) = 10.5 \text{ m/s}^2$$

EXAMPLE

Two boxes are attached by a massless rope that hangs over a pulley. The masses of the boxes are 10 kg and 20 kg. The pulley has a radius of 0.31 m and a mass of 16 kg. What is the acceleration of each box?

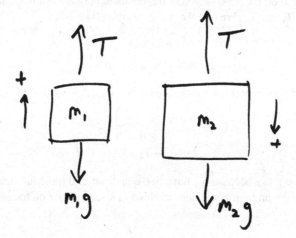

We did this before when we didn't have to include the rotation of the pulley and got

$$\begin{cases} +T - m_1 g = m_1 a \\ +m_2 g - T = m_2 a \end{cases}$$

Now the pulley rotates as the rope moves over it. The force that causes the pulley to spin is friction between the rope and the pulley.

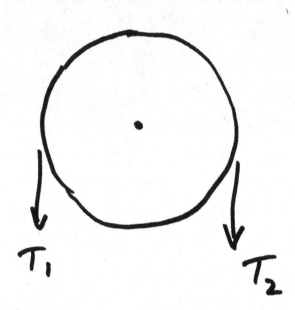

We can use the two tensions to calculate the torque on the pulley. Because there is another force on the rope between the tensions, they are not necessarily equal, so we call them T_1 and T_2. We add the forces on the 10 kg box:

$$F = ma$$
$$+T_1 - m_1 g = m_1 a$$

Can we solve this to find the acceleration? No, because we don't know the tension. So let's try the 20 kg box:

$$F = ma$$
$$+m_2 g - T_2 = m_2 a$$

We can't solve these because we have two equations and three unknowns.

We use the angular equivalent to Newton's second law on the pulley:

$$\tau = I\alpha$$
$$+T_2 R - T_1 R = \left(\frac{1}{2} m_P R^2\right)\alpha$$

At the point that each rope acts on the pulley it is perpendicular to the pulley, so the angle is 90°. We now have another equation, but we introduced another unknown (α).

We can connect the angular acceleration with the linear acceleration using if $a = R\alpha$, so

$$(T_2 - T_1) R = \left(\frac{1}{2}m_P R^2\right)\frac{a}{R}$$

Now we have a set of equations we can solve

$$\begin{cases} +T_1 - m_1 g = m_1 a \\ + m_2 g - T_2 = m_2 a \\ + T_2 - T_1 = \frac{1}{2}m_P a \end{cases}$$

Adding the equations

$$m_2 g - m_1 g = \left(m_1 + m_2 + \frac{1}{2}m_P\right) a$$

$$a = \frac{(m_2 - m_1)g}{(m_1 + m_2 + \frac{1}{2}m_P)} = \frac{(20\text{ kg} - 10\text{ kg})(9.8\text{ m/s}^2)}{(10\text{ kg} + 20\text{ kg} + \frac{1}{2}16\text{ kg})} = 2.6\text{ m/s}^2$$

Why did the radius of the pulley disappear from the equations? The greater the radius, the greater the moment of inertia, but the greater the torque that the same tension applies to the pulley.

EXAMPLE

A 6.5 kg (M_{board}), 2.0 m long (L) board is supported by a 8 cm high (h) pivot at its middle. It tips when a 1.0 kg mass (M_1) is placed on the board 10 cm (x) from the pivot. How long does it take for the end of the board to hit the ground?

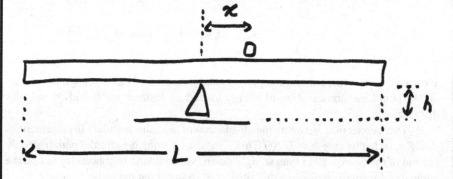

Since the board is pivoting we need to use rotation. The 1.0 kg mass will cause a torque on the board that is nearly constant, so the angular acceleration will be nearly constant.

$$F = ma \rightarrow \tau = I\alpha$$

If we can find the torque τ and the moment of inertia I, then we can find the angular acceleration α. With α we can try our constant acceleration technique.

The forces acting on the board are the weight of the board, the normal force from the pivot, and the weight of the 1.0 kg mass (really the normal force from the mass, but if the acceleration of the mass is small then it's nearly equal to the weight of the mass). The first two forces act at the pivot point, so their moment arms are zero and they don't create any torque on the board.

$$\tau = FR \sin \phi$$

$$\tau = (M_1 g)(x) \sin 90° = M_1 g x$$

The moment of inertia (angular mass) of the board and the 1.0 kg mass is

$$I = I_{board} + I_1$$

$$I = \frac{1}{12} M_{board} L^2 + M_1 x^2$$

The angular acceleration is

$$\tau = I\alpha$$

$$M_1 g x = \left(\frac{1}{12} M_{board} L^2 + M_1 x^2 \right) \alpha$$

$$\alpha = \frac{M_1 g x}{\frac{1}{12} M_{board} L^2 + M_1 x^2}$$

With the angular acceleration, we try the constant acceleration technique.

Δx	$\rightarrow$	$\Delta\theta$	=	
v_0	$\rightarrow$	ω_0	=	0
v	$\rightarrow$	ω	=	
a	$\rightarrow$	α	=	$(M_1 g x)/\left(\frac{1}{12} M_{board} L^2 + M_1 x^2\right)$
t	$\rightarrow$	t	=	?

We need a third value, either the angular displacement $\Delta\theta$ or the final angular velocity ω. We could use conservation of energy to find ω. Instead, we'll find the angular displacement.

The connection between the displacement and the angular displacement is $x = R\theta$, or in this case $h = L \, \Delta\theta$. This is measured along the circular path traced by the end of the board, rather than straight down. We could use trigonometry to find the angle, but for small angles like this $\sin\theta \approx \theta$, so it's about the same.

We don't care about the final angular velocity, so we use

$$\Delta x = v_0 t + \frac{1}{2} a t^2 \quad \rightarrow \quad \Delta\theta = \omega_0 t + \frac{1}{2} \alpha t^2$$

$$\frac{h}{L} = (0)t + \frac{1}{2} \left(\frac{M_1 g x}{\frac{1}{12} M_{board} L^2 + M_1 x^2} \right) t^2$$

$$t = \sqrt{\frac{2h\left(\frac{1}{12}M_{board}L^2 + M_1x^2\right)}{LM_1gx}}$$

$$t = \sqrt{\frac{2(0.08 \text{ m})\left[\frac{1}{12}(6.5 \text{ kg})(2.0 \text{ m})^2 + (1.0 \text{ kg})(0.10 \text{ m})^2\right]}{(2.0 \text{ m})(1.0 \text{ kg})(9.8 \text{ m/s}^2)(0.10 \text{ m})}}$$

$$t = 0.42 \text{ s}$$

7.4 ENERGY AND ROLLING MOTION

Everything we did with energy works with angular quantities:

$$KE = \tfrac{1}{2}mv^2 \qquad \rightarrow \qquad KE = \tfrac{1}{2}I\omega^2$$
$$W = Fd \qquad \rightarrow \qquad W = \tau\,\Delta\theta$$

EXAMPLE

A boy leaving a store pushes on the door handle with a force of 18 N. The door is 0.78 m wide and has a mass of 7.2 kg. The boy pushes perpendicular to the surface of the door. How fast is the door rotating when it opens (i.e., rotates to 90°)?

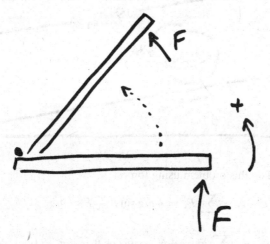

We could do this using constant angular acceleration, especially since we found most of the relevant intermediate quantities earlier. We could also use conservation of energy. A clue that energy is worth trying is that neither an acceleration nor a time is involved.

$$KE + PE + W = KE' + PE'$$

$$(0) + (0) + \tau\,\Delta\theta = \frac{1}{2}I(\omega')^2 + (0)$$

We found in an earlier example that the torque is

$$\tau = FL\sin 90° = 14\ \text{N m}$$

and that the moment of inertia is

$$I = \frac{1}{3}ML^2 = \frac{1}{3}(7.2\ \text{kg})(0.78\ \text{m})^2 = 1.46\ \text{kg m}^2$$

$$(14\ \text{N m})\left(\frac{\pi}{2}\right) = \frac{1}{2}(1.46\ \text{kg m}^2)(\omega')^2$$

$$\omega' = 5.5\ \text{rad/s}$$

Rotating motion does not usually have potential energy. It could—imagine a trapdoor into an attic being opened—but in many common situations it doesn't.

EXAMPLE

A 7.0 kg bowling ball rolls without slipping down a ramp. The ramp is 1.4 m high and 9.6 m long. How fast is the bowling ball moving when it reaches the bottom of the ramp?

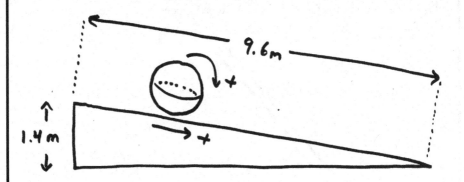

We could do this problem using forces.

$$\sum F_x = +mg\sin\beta - \mathcal{F} = ma_x$$

$$\sum F_y = +N - mg\cos\beta = ma_y = 0$$

Here, β is the angle the ramp makes with the floor. We don't know the friction force $\mathcal{F}$ or the acceleration a_x. Because the ball rolls without slipping, the friction is static friction and we can't use $\mathcal{F} = \mu N$. We need another piece of information. That piece is the rotation

$$\sum \tau = \mathcal{F}R \sin 90° = I\alpha$$

Gravity and the normal force do not create any torque about the center of the ball. We connect α and a_x using

$$a_x = R\alpha$$

When we put all of those together, we find the acceleration of the ball down the ramp. We then use the constant acceleration technique to find the final velocity.

Alternatively, we could use conservation of energy. The fact that we want the final speed and don't care about the acceleration or the time is a tip-off that energy is a good place to start.

$$KE + PE + W = KE' + PE'$$

$$(0) + mgh + W = \left(\frac{1}{2}m(v')^2 + \frac{1}{2}I(\omega')^2\right) + mg(0)$$

We have kinetic energy both from moving and from rolling. The work done by the normal force is zero because the force is perpendicular to the motion. The work done by the friction force is zero because there is no sliding, so there is no motion at the point of contact.

$$mgh = \frac{1}{2}m(v')^2 + \frac{1}{2}I(\omega')^2$$

We have two unknowns and only one equation. But we can connect the rotating and linear motion with $v = R\omega$, so

$$mgh = \frac{1}{2}m(v')^2 + \frac{1}{2}\left(\frac{2}{5}mR^2\right)\left(\frac{v'}{R}\right)^2$$

$$gh = \frac{7}{10}(v')^2$$

$$v' = \sqrt{\frac{10}{7}gh}$$

$$v' = \sqrt{\frac{10}{7}(9.8 \text{ m/s}^2)(1.4 \text{ m})}$$

$$v' = 4.4 \text{ m/s}$$

Notice how both m and R cancel from the equation. It doesn't matter what the size or mass of the ball is, it will have the same speed when it gets to the bottom (the mass didn't matter if it slid without friction). What does matter is the shape—if we used a hollow sphere or a tube instead of a solid sphere, the 2/5 would have been something else, and we would have gotten a different result.

7.5 ANGULAR MOMENTUM AND ANGULAR IMPULSE

Everything we did with momentum works with angular quantities:

$$P = mv \qquad \rightarrow \qquad L = I\omega = PR\sin\phi$$
$$\Delta P = Ft \qquad \rightarrow \qquad \Delta L = \tau t$$

For a rotating object, we use $I\omega$, but for a point object moving around a point we typically use $PR\sin\phi$. Angular impulse is the name given to the change in the angular momentum, which thus has the same units as angular momentum ($kg\,m^2/s$). The basic equation is

$$L_{before} + \Delta L = L_{after} \quad \text{or} \quad L + \Delta L = L'$$

EXAMPLE

An ice skater doing an axel pulls her arms in to spin faster. If she triples her rotational speed, what does she do to her moment of inertia?

Since there is nothing in the problem about acceleration or time, a conservation law is a good place to start. It takes force for the skater to pull in her arms, so she does work in the process. Since we don't know how much work she does, conservation of energy looks unpromising. She is rotating rather than moving in a straight line, so let's try conservation of angular momentum.

$$L + \Delta L = L'$$

The force that she exerts to pull her arms in is directly toward her, so the force is antiparallel to the moment arm and the torque is zero. Since the torque is zero, the angular impulse is zero.

$$I\omega + (0) = I'\omega'$$

$$I\omega = I'(3\omega)$$

$$I' = \frac{1}{3}I$$

By pulling in her arms she decreases her moment of inertia by a factor of 3.

What does this do to her kinetic energy?

$$KE' = \frac{1}{2}I'(\omega')^2 = \frac{1}{2}(\frac{1}{3}I)(3\omega)^2 = 3\frac{1}{2}I\omega^2 = 3KE$$

Where does this extra energy come from? She exerts an inward force on her arms as they move inward, so she does positive work and her energy increases.

EXAMPLE

A small object moves in a circle on the end of a string on a horizontal frictionless table. Initially the object moves at 0.74 m/s at the end of a 17 cm string. The string is then pulled through a hole in the center of the table. When the string remaining is only 5.0 cm, how fast is the object moving?

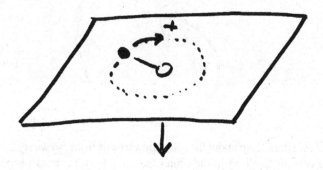

Again, we don't care about the acceleration or the time, so we try a conservation law.

$$L + \Delta L = L'$$

The force exerted on the object by the string is antiparallel to the moment arm and creates no torque. The normal force cancels the gravity, so the total torque is zero. We use $L = PR\sin\phi$ for the angular momentum.

$$PR\sin\phi + (0) = P'R'\sin\phi'$$

The object is moving in a circle, with the momentum perpendicular to the moment arm, both before and after the string is pulled, so $\phi = \phi' = 90°$.

$$PR = P'R'$$

$$mvR = mv'R'$$

$$v' = \frac{vR}{R'}$$

$$v' = \frac{(0.74 \text{ m/s})(17 \text{ cm})}{(5.0 \text{ cm})}$$

$$v' = 2.5 \text{ m/s}$$

EXAMPLE

A roller coaster enters a spiral in which the radius of the track decreases from 17 m to 5.0 m while staying horizontal. If the coaster enters the spiral at a speed of 7.4 m/s, what is its speed at the end of the spiral?

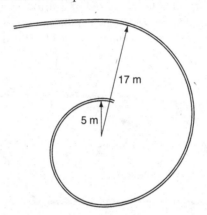

Again, we don't care about the acceleration or the time, so we try a conservation law. As the cars move closer to the center the force from the tracks is perpendicular to the tracks (it's a normal force) but *not* directed toward the center of the spiral. The force is not antiparallel to the moment arm, the torque is not zero, and the angular impulse is not zero. Therefore, conservation of angular momentum, which worked so well in the last example, doesn't work here—it's still true, but since we can't calculate the force from the tracks (at least not easily), we can't use it to get our answer.

If the force is perpendicular to the motion, then the work is zero, and conservation of energy looks promising.

$$KE + PE + W = KE' + PE'$$

$$\frac{1}{2}mv^2 + (0) + (0) = \frac{1}{2}m(v')^2 + (0)$$

$$v' = v = 7.4 \text{ m/s}$$

Since no work is done on the roller coaster, the kinetic energy doesn't change and the speed is unchanged.

The difference between the last example and this one is the direction of the force. In the last example it was always toward the pivot point, but in this example it was not.

EXAMPLE

A 0.85 kg, 35 cm long uniform rod falls from vertical to horizontal, pivoting about its end, then strikes a 0.56 kg block of putty. The block sticks to the end of the rod. What is the speed of the putty after it is hit by the rod?

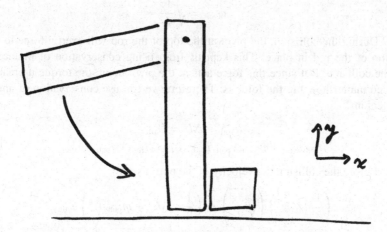

This problem appeared in the last chapter and we couldn't solve it (yet). The first difficulty was that not all of the rod will be moving with the same speed. We can now describe the motion of the rod using rotation. The second difficulty was dealing with the impulse from the pivot during the collision. We now have the tool we need to avoid that.

Following the solution from the example in the last chapter, we want to use conservation of energy to find the speed of the rod at the time of the collision.

$$KE_{top} + PE_{top} + W = KE_{vert} + PE_{vert}$$

$$\frac{1}{2}I_{rod}(0)^2 + m_{rod}gh_{top} + W = \frac{1}{2}I_{rod}(\omega_{vert})^2 + m_{rod}gh_{vert}$$

The work done by the pivot is zero because the pivot doesn't move and there is no friction in the rotation about the pivot. The work done by gravity is taken care of with potential energy.

$$\frac{1}{2}I_{rod}(\omega_{vert})^2 = m_{rod}gh_{top} - m_{rod}gh_{vert}$$

Using the parallel axis theorem, the moment of inertia of the rod is

$$I_{rod} = \frac{1}{12}m_{rod}L^2 + m_{rod}(L/2)^2 = \frac{1}{3}m_{rod}L^2$$

The change in the height of the rod is the change in the height of the center, which is half the length.

$$\frac{1}{2}\left(\frac{1}{3}m_{rod}L^2\right)(\omega_{vert})^2 = m_{rod}g\left(\frac{1}{2}L\right)$$

$$\frac{1}{3}L(\omega_{vert})^2 = g$$

$$\omega_{vert} = \sqrt{\frac{3g}{L}}$$

During the collision, the pivot at the top of the rod will exert a force to keep the top of the rod in place. This kept us from using conservation of momentum for the collision. But since this force acts at the pivot point, the torque it creates is zero no matter how big the force is. Therefore we can use conservation of angular momentum.

$$L_{vert} + \Delta L = L_{after}$$
$$I_{rod}\omega_{vert} + m_{putty}v_{putty}L + (0) = (I_{rod} + I_{putty})\omega_{after}$$

Before the collision the velocity of the putty is zero.

$$\left(\frac{1}{3}m_{rod}L^2\right)\left(\sqrt{\frac{3g}{L}}\right) = \left(\frac{1}{3}m_{rod}L^2 + m_{putty}L^2\right)\omega_{after}$$

$$\omega_{after} = \frac{m_{rod}}{m_{rod} + 3m_{putty}}\sqrt{\frac{3g}{L}}$$

The speed of the putty is

$$v = R\omega = L\omega_{after}$$

$$v = \frac{m_{rod}}{m_{rod} + 3m_{putty}}\sqrt{3gL}$$

$$v = \frac{(0.85\text{ kg})}{(0.85\text{ kg}) + 3(0.56\text{ kg})}\sqrt{3(9.8\text{ m/s}^2)(0.35\text{ m})}$$

$$v = 1.08\text{ m/s}$$

LESSON

Conservation of angular momentum lets you ignore forces at the pivot point or toward the pivot point.

7.6 STATICS

Static means unchanging, and in physics (and engineering) statics is the study of objects that aren't moving. By this we mean that not only is an object not moving

($v = 0$), but its motion isn't changing ($a = 0$); not only is it not rotating ($\omega = 0$), but its rotation isn't changing ($\alpha = 0$).

EXAMPLE

A 2.0 m long, 6.0 kg board is supported by two scales, one at the left end and the other 1.2 m from the left end. What is the reading on each scale?

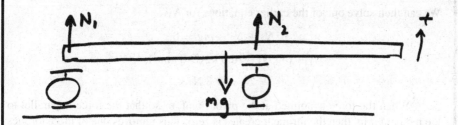

Each scale pushes up on the board while measuring the force: We add up the forces:

$$F = ma$$

$$0 \text{ (none)} = ma_x = 0$$

$$N_1 + N_2 - mg = ma_y = 0$$

The first equation does us no good—there are neither forces nor motion in the horizontal direction. We can't solve the second equation (yet) because there are two unknowns and only one equation. We need another piece of information.

Since the board continues to not rotate, the angular acceleration is zero. Since it is not rotating about any point, the angular acceleration about any point is zero, and thus the torque about any point is zero. Let's use the center of the board as our pivot point, and take counterclockwise as the positive direction for rotation:

$$\sum \tau = I\alpha$$

$$\sum_i F_i R_i \sin \phi_i = 0$$

$$-N_1(1.0 \text{ m})(1) + N_2(0.2 \text{ m}) \underbrace{(1)}_{\sin 90°} +(mg)(0) \sin(?) = 0$$

$$-N_1(1.0) + N_2(0.2) = 0$$

We now have two equations with two unknowns, so we can solve them (note that the torque caused by N_1 is in our negative direction).

Alternatively, what if we had chosen the left end as our pivot point?

$$\sum_i F_i R_i \sin \phi_i = I\alpha = 0$$

$$N_1(0)\sin(?) + N_2(1.2 \text{ m}) \underbrace{(1)}_{\sin 90°} - (mg)(1.0 \text{ m})(1) = 0$$

$$+N_2(1.2) - (mg)(1.0) = 0$$

We can solve this immediately for N_2.

$$N_2 = (mg)(1.0)/(1.2) = \frac{1.0}{1.2}(6.0 \text{ kg})(9.8 \text{ m/s}^2) = 49 \text{ N}$$

We can then solve one of the earlier equations for N_1.

$$N_1 + N_2 - mg = 0$$

$$N_1 + (49 \text{ N}) - (6.0 \text{ kg})(9.8 \text{ m/s}^2) = 0$$

$$N_1 = 9.8 \text{ N}$$

When the force is applied at the pivot point, or so that the force is parallel to the moment arm, then the torque it creates is zero. This removes one or more torques from the equation and makes it easier to solve. This is especially helpful when the eliminated torque was created by one of the unknown forces.

EXAMPLE

A 75 kg diver (165 lb) stands at the end of a diving board. The nearly massless board is 3.7 m long and has two supports, 0.90 m apart, with one at the far end from the diver. What is the force from each support on the diving board?

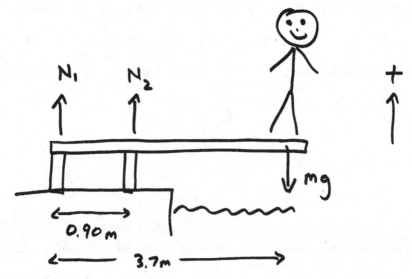

This problem is very similar to the last one. We add the vertical forces

$$\sum F_y = ma_y$$

$$N_1 + N_2 - mg = 0$$

It is not true that the weight of the diver acts on the diving board. The weight of the diver acts on the diver, and the normal force from the board also acts on the diver. Since these add to zero (so that $a_{diver} = 0$), the normal force of the board on the diver, and thus the force of the diver on the board, is equal to the weight of the diver. Having said that, we won't mention this step from now on. Since N_1 is unknown we add the torques about the left end.

$$\sum \tau = I\alpha$$

$$\sum_i F_i R_i \sin \phi_i = 0$$

$$N_1(0) \sin(?) - N_2(0.90 \text{ m}) \underbrace{(1)}_{\sin 90°} + (mg)(3.7 \text{ m})(1) = 0$$

$$-N_2(0.90) + (mg)(3.7) = 0$$

$$N_2 = \frac{(mg)(3.7)}{(0.90)} = \frac{3.7}{0.90}(75 \text{ kg})(9.8 \text{ m/s}^2) = 3020 \text{ N}$$

$$N_1 + N_2 - mg = 0$$

$$N_1 + (3020 \text{ N}) - (75 \text{ kg})(9.8 \text{ m/s}^2) = 0$$

$$N_1 = -2290 \text{ N}$$

Can a normal force be negative? The negative sign means that the left support must pull the left end of the board down. This can only happen if the board is attached to the left support rather than just resting on it. Does this make sense? What would have happened if we had chosen the midpoint of the board, or the right support, for our pivot point? If the left support pushed up, then all of the torques would have been clockwise, and the total torque could not have been zero.

EXAMPLE

A 25 kg sign is hanging from a 3.0 m long, 14 kg beam. The end of the beam is supported by a wire that makes a 39° angle with the beam. What is the tension in the wire?

We not only don't know the force that the wall puts on the beam, we don't know the direction of the force that the wall puts on the beam. But whatever its direction, it has an x component and a y component and we can use these. We add up the forces:

$$F = ma$$

$$+ W_x - T \cos 39° = m_{beam} a_x = 0$$

$$+ W_y - M_{beam} g - M_{sign} g + T \sin 39° = m_{beam} a_y = 0$$

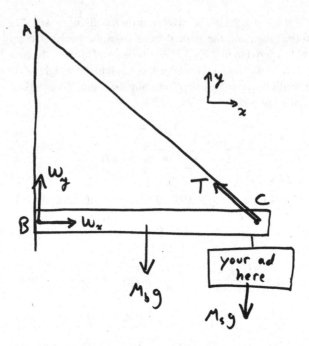

Since we don't know W_x, W_y, or T, we can't solve either of these yet. We need to add up torques. By choosing the pivot point wisely, we can eliminate two of the torques from the equation (we could eliminate only one before because all of the forces were parallel). Because our unknown forces are W_x, W_y, and T, we would like to eliminate two of these. To do this, we choose one of A, B, or C as our pivot point. Can we use A, even though it is not on the object? Sure! (We could choose Paris as our pivot point and it would work, but the choice of Paris doesn't make the equations easier to solve, so we won't.)

Since we want to find the tension T, we'll use B as the pivot point, eliminating W_x and W_y and giving us an equation we can solve for T.

$$\sum_i F_i R_i \sin \phi_i = I\alpha = 0$$

$$W_x(0) \sin(?) + W_y(0) \sin(?) - M_{\text{beam}}g\left(\frac{1}{2}L\right)(1) - M_{\text{sign}}g(L)(1) + T(L) \sin 39° = 0$$

$$-\frac{1}{2}M_{\text{beam}}g - M_{\text{sign}}g + T \sin 39° = 0$$

$$T = \left(\frac{1}{2}M_{\text{beam}}g + M_{\text{sign}}g\right) / \sin 39°$$

$$T = \left(\frac{1}{2}(14 \text{ kg}) + (25 \text{ kg})\right)(9.8 \text{ m/s}^2) / \sin 39°$$

$$T = 498 \text{ N}$$

LESSON

In a statics problem, it is always possible to eliminate at least one torque from the torque equation by wise choice of the pivot point. Choose to eliminate a torque caused by an unknown force.

EXAMPLE

A ladder of mass M and length L leans against a smooth building and has a coefficient of friction μ with the ground. A man of mass m stands a distance x up the ladder. At what angle θ with the ground will the ladder begin to slip? To prevent slipping, should the angle be greater or smaller than this angle?

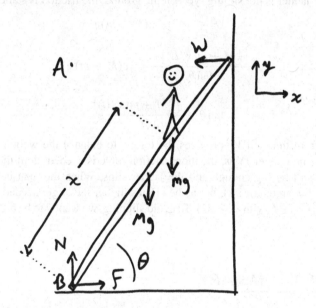

We begin by adding up the forces:

$$+\mathcal{F} - \mathcal{W} = Ma_x = 0$$
$$+N - Mg - mg = Ma_y = 0$$

The "smooth" building means we don't include the friction between the ladder and the building.

We don't know $\mathcal{F}$, $\mathcal{W}$, or N, and we want to find $\mathcal{F}$, so it might make sense to choose A as our pivot point. This would make writing the torque equation much more difficult (try imagining the moment arms) so we'll use B instead.

$$\sum_i F_i R_i \sin \phi_i = I\alpha = 0$$

$$N(0)\sin(?) + \mathcal{F}(0)\sin(?) - Mg\left(\frac{1}{2}L\right)\sin(90° - \theta) - mg(x)\sin(90° - \theta)$$
$$+\mathcal{W}(L)\sin\theta = 0$$
$$-Mg\left(\frac{1}{2}L\right)\cos\theta - mg(x)\cos\theta + \mathcal{W}(L)\sin\theta = 0$$

Here we've used $\sin(90° - \theta) = \cos\theta$.

$$\mathcal{W} = \frac{\left(\frac{1}{2}ML + mx\right)g}{L}\frac{\cos\theta}{\sin\theta}$$

$$\mathcal{F} = \mathcal{W} = \frac{\left[\frac{1}{2}M + m(x/L)\right]g}{\tan\theta}$$

Because the ladder is not sliding (yet) on the ground, the friction is static friction:

$$\mathcal{F} \leq \mu N = \mu(M + m)g$$

$$\frac{\left[\frac{1}{2}M + m(x/L)\right]g}{\tan\theta} \leq \mu(M + m)g$$

$$\tan\theta \geq \frac{\left[\frac{1}{2}M + m(x/L)\right]}{\mu(M + m)}$$

The force from the wall $\mathcal{W}$ provides the torque to balance the weight of the man and ladder. The greater $\mathcal{W}$ is, the more friction force is needed, until the static friction force can't be big enough and the ladder slips. When the man is halfway up the ladder, $\theta = \arctan(1/2\mu)$, but as he climbs higher the angle needed approaches $\arctan(1/\mu)$ $(x \to L$ with $m \geq M)$. To avoid slipping, we want θ to be big and μ to be big.

CHAPTER SUMMARY

- Use rotation for objects that have a fixed pivot point, even if they aren't rotating at the time.
- Everything we've done so far still works, but with angular quantities:

$$\begin{aligned}
x &\to \theta \\
v &\to \omega \\
a &\to \alpha \\
m &\to I \\
F &\to \tau \\
P &\to L \\
\Delta x = v_0 t + \tfrac{1}{2}at^2 &\to \Delta\theta = \omega_0 t + \tfrac{1}{2}\alpha t^2 \\
F = ma &\to \tau = I\alpha \\
P = mv &\to L = I\omega
\end{aligned}$$

- We connect the linear (tangential) and angular quantities with

$$x = R\theta$$
$$v = R\omega$$
$$a = R\alpha$$
$$\tau = FR\sin\phi$$
$$L = PR\sin\phi$$

- Watch out for R (or r)! R is always the distance to the pivot point, but it can be used multiple times in the same problem to refer to different distances.
- In a statics problem, choose the pivot point to eliminate the torque from a force you don't know.

CHAPTER *8*

ELECTRIC AND MAGNETIC FIELDS

Electricity is all about charges, how they move, and getting them to do work. What is charge? It is a property of some particles. The important characteristic of a charged particle is that it creates forces on other charged particles. Charged particles create electric fields, which create forces on charged particles. Moving charged particles create magnetic fields, which create forces on moving charged particles. What is a "field"? We'll get to that soon.

The electric and magnetic forces are just like any other force, so all of the techniques we've learned still work. When a charge experiences an electric force, we draw a free body diagram of all the forces including the electric force, just like before. Electric and magnetic forces are like gravity in that they work at a distance; two charges do not have to be touching to create a force, in the same way that Earth does not have to be touching an apple to pull down on it with gravity.

Here are five things to remember about the behavior of charges:

- Charge comes in two types, which we call positive and negative.
- Opposite charges attract and like charges repel.
- The closer the charges are to one another, the stronger the force.
- Charges can move around in a conducting material, but not in an insulating material.
- "Ground" can supply an infinite amount of charge.

When a positive and a negative charge create forces on each other, the forces are not necessarily in the negative direction. From Newton's third law, the force on one is in the opposite direction as the force on the other.

EXAMPLE

A positively charged rod is brought near to a neutral conductor. What is the direction of the force on the conductor?

A neutral object is one that has no net charge; it has an equal number of protons (+) and electrons (−). But the charges don't have to be evenly distributed. When a positively charged object is brought near, the negative electrons are attracted to it. Because they are in conducting material, they can move around, so some of them move toward the rod.

Likewise the positive charges in the conductor are repelled by the rod and move away from it. (Really it is only the negative charges that move, and they leave behind

160

an excess of positive charges, but it's easier to say and think about positive charges also moving.) When charge moves within a conductor due to nearby charges we call this "induced charge."

The negative charges are closer to the rod than the positive charges, so the electrical force on them is stronger. The net force on the conductor is toward the rod. Likewise the net force on the rod is toward the conductor.

EXAMPLE ·

A positively charged rod is brought near to a grounded conductor. The connection to ground is removed and then the rod is taken away. What is the charge on the conductor? What if the order were reversed—the rod is taken away and then the connection to ground removed?

When we say "ground" we mean an object so big that we can take charge out of it and it will still be neutral, or so close to neutral that no one can tell. When we bring the positively charged rod near to the conductor, negative charges are attracted to the rod. Some negative charges (electrons) come out of ground onto the conductor, so that it has a net negative charge.

If the connection to ground is removed, the negative charges no longer have a path off of the conductor—they are trapped. When the rod is taken away the conductor still has a negative charge on it.

What if the connection to ground had still been there when the rod was taken away? The electrons would have had a path to leave the conductor. Since the electrons repel each other, and try to get as far from one another as possible, they would have gone back into ground so that they could spread out further. The conductor is then neutral.

LESSON

Charges can move around in conducting material. Grounded conductors are not necessarily neutral—they can have any charge.

8.1 ELECTRIC FORCES

Any two charged objects create a force on each other, in the same way that any two masses create a force on each other. The difference is that charges come both positive and negative, so that the electric force can be either attractive or repulsive. The magnitude of the electric force two charges create on each other is

$$F_Q = \frac{kq_1q_2}{R^2}$$

where $k = 9 \times 10^9$ N m^2/C^2, q_1 and q_2 are the charges, and R is the distance between them. This is known as Coulomb's law and the force is known as a "coulomb force." Charge is typically represented by either q or Q, but if both occur in the same problem they refer to two different charges.

Charge is measured in coulombs (C), with the charge on a single proton equal to

$$e = 1.6 \times 10^{-19} \text{ C}$$

so that a coulomb is about 6×10^{18} protons or electrons. The charge on an electron is $-e$. Like with g, e is not negative; it is the magnitude of the charge on the proton or electron.

We treat electric forces like we would any other force. Draw the free body diagram, choose axes, and add the forces. Before there was only one force that could act without being in contact with the object: gravity. Now we have electric forces that act without being in contact.

EXAMPLE

Find the total electric force on the middle charge ($+1$ μC).

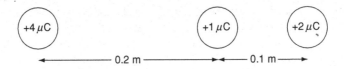

The $+4$ μC charge and the $+1$ μC charge have the same sign, so they repel each other. The $+4$ μC charge creates a force on the $+1$ μC charge that is away from the $+4$ μC charge, or to the right. The magnitude or size of this force is

$$F_4 = \frac{kq_1q_2}{d^2} = \frac{k(4\ \mu C)(1\ \mu C)}{(0.2\ \text{m})^2}$$

$$F_4 = \frac{(9 \times 10^9\ \text{N m}^2/\text{C}^2)(4 \times 10^{-6}\ \text{C})(1 \times 10^{-6}\ \text{C})}{(0.2\ \text{m})^2}$$

$$F_4 = 0.9\ \text{N}$$

The $+2$ μC charge and the $+1$ μC charge have the same sign, so they also repel each other. The $+2$ μC charge creates a force on the $+1$ μC charge that is away from the $+2$ μC charge, or to the left. The magnitude or size of this force is

$$F_2 = \frac{kq_1q_2}{d^2} = \frac{k(2\ \mu C)(1\ \mu C)}{(0.1\ \text{m})^2}$$

$$F_2 = \frac{(9 \times 10^9\ \text{N m}^2/\text{C}^2)(2 \times 10^{-6}\ \text{C})(1 \times 10^{-6}\ \text{C})}{(0.1\ \text{m})^2}$$

$$F_2 = 1.8\ \text{N}$$

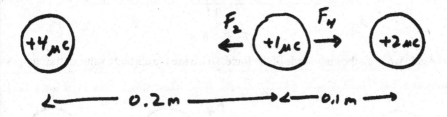

We can add the two forces using right as the positive direction. The total electric force on the $+1$ μC charge is

$$F = +F_4 - F_2 = +(0.9\ \text{N}) - (1.8\ \text{N}) = -0.9\ \text{N}$$

or 0.9 N to the left. Even though the left charge is twice as big (twice as powerful, twice as electric?), it's also twice as far away and creates less force on the center charge.

EXAMPLE

Find the total electric force on the middle charge $(-1\ \mu C)$.

The $+4\ \mu C$ charge and the $-1\ \mu C$ charge have opposite signs, so they attract each other. The $+4\ \mu C$ charge creates a force on the $-1\ \mu C$ charge that is toward the $+4\ \mu C$ charge, or to the left. The magnitude or size of this force is

$$F_4 = \frac{kq_1q_2}{d^2} = \frac{k(4\ \mu C)(1\ \mu C)}{(0.2\ m)^2}$$

$$F_4 = \frac{(9 \times 10^9\ N\ m^2/C^2)(4 \times 10^{-6}\ C)(1 \times 10^{-6}\ C)}{(0.2\ m)^2}$$

$$F_4 = 0.9\ N$$

Why don't we include the sign of the $-1\ \mu C$ charge in the equation? Because the formula tells us the *magnitude* of the force, and we use the diagram to figure out the direction.

The $-2\ \mu C$ charge and the $-1\ \mu C$ charge have the same sign, so they repel each other. The $-2\ \mu C$ charge creates a force on the $-1\ \mu C$ charge that is away from the $-2\ \mu C$ charge, or to the left. The magnitude or size of this force is

$$F_2 = \frac{kq_1q_2}{d^2} = \frac{k(2\ \mu C)(1\ \mu C)}{(0.1\ m)^2}$$

$$F_2 = \frac{(9 \times 10^9\ N\ m^2/C^2)(2 \times 10^{-6}\ C)(1 \times 10^{-6}\ C)}{(0.1\ m)^2}$$

$$F_2 = 1.8\ N$$

Again we want the magnitude of the force, so we use the absolute value of the charges.

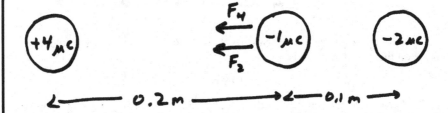

Using right as the positive direction, the total electrical force on the $-1\ \mu C$ charge is

$$F = -F_4 - F_2 = -(0.9\ N) - (1.8\ N) = -2.7\ N$$

or 2.7 N to the left.

EXAMPLE

Three $+2\ \mu C$ charges lie at the corners of an equilateral triangle. The length of each side of the triangle is 2 cm. Find the electric force on one of the charges.

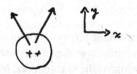

Because all of the charges are positive, they all repel one another. We can find the force on any of the charges because of the symmetry—rotate the figure 120° and everything looks the same. The drawing shows the forces on two of the charges and possible coordinate systems.

Find the forces on the top charge, in both the x and y directions.

$$F_Q = \frac{kq_1q_2}{d^2} = \frac{(9 \times 10^9 \text{ N m}^2/\text{C}^2)(2 \ \mu\text{C})(2 \ \mu\text{C})}{(0.02 \text{ m})^2} = 90 \text{ N}$$
$$F_x = +F_Q \cos 60° - F_Q \cos 60° = 0$$
$$F_y = +F_Q \sin 60° + F_Q \sin 60°$$
$$F_y = 2F_Q \sin 60° = 2(90 \text{ N}) \sin 60° = 156 \text{ N}$$

The force on the top charge is 156 N upward.

Now find the forces on the bottom right charge, in both the x and y directions.

$$F_x = +F_Q + F_Q \cos 60° = 135 \text{ N}$$
$$F_y = +F_Q \sin 60° = 78 \text{ N}$$
$$F = \sqrt{F_x^2 + F_y^2} = \sqrt{(135 \text{ N})^2 + (78 \text{ N})^2} = 156 \text{ N}$$

The force on the bottom right charge is also 156 N.

LESSON

Use the formula to find the magnitude and the diagram to find the direction of electric forces, just like any other force.

EXAMPLE

An electroscope consists of two light pieces of metal that, when charged, repel each other. Treat the mass of the electroscope as being entirely at the ends of the "leaves," which are $L = 6$ cm long. Determine the mass needed for each leaf so that a $Q = 4\,\mu C$ charge causes the leaves to stand up $40°$.

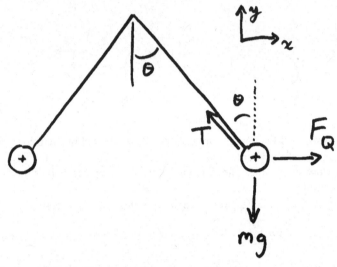

We start with the free body diagram: each leaf has mass and charge, so there are gravity and electric forces. There is also a tension T along the leaf. There is no acceleration, so we pick our axes to make dividing the forces into components easier.

$$F = ma$$
$$+F_Q - T \sin 40° = ma_x = 0$$
$$-mg + T \cos 40° = ma_y = 0$$

We want to solve for the mass, but we don't know the tension. We could solve the x equation for the tension if we knew the electric force. We can find the electric force from the charges. The charge on the electroscope splits with half going to each leaf.

$$F_Q = \frac{k(Q/2)^2}{R^2} = \frac{k(Q/2)^2}{(2L \sin 40°)^2}$$

$$T = \frac{F_Q}{\sin 40°} = \frac{kQ^2}{4 \times 4L^2 (\sin 40°)^3}$$

$$mg = T \cos 40° = \frac{kQ^2}{16L^2} \frac{\cos 40°}{(\sin 40°)^3}$$

$$m = \frac{(9 \times 10^9 \text{ N m}^2/\text{C}^2)(4\,\mu C)^2}{16(9.8 \text{ m/s}^2)(0.06 \text{ m})^2} \frac{\cos 40°}{(\sin 40°)^3} = 0.74 \text{ kg}$$

EXAMPLE ★

A charge of $+Q$ is spread out uniformly over a rod of length L. A second charge q is along the line of the rod a distance d from the nearest end. What is the force on the charge q?

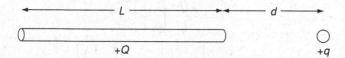

We would like to use Coulomb's law, but what is the distance between the two charges q and Q? We could measure from the middle of the rod, but $1/R^2$ is nonlinear so that won't work.

We could divide the rod in half, find the force from each half, and add them (as vectors, but both halves create a force to the right).

$$F = \frac{k(Q/2)q}{[d+(L/4)]^2} + \frac{k(Q/2)q}{[d+(3L/4)]^2}$$

This would be closer. We could divide the rod into three pieces for a better answer, or four or more. We want to divide the rod into pieces so small that we can say that all of the charge in each piece, is the same distance from the charge q. That is, we want to divide the rod into an infinite number of infinitely small pieces, find the force from each piece, and add the forces. This sounds like an integral.

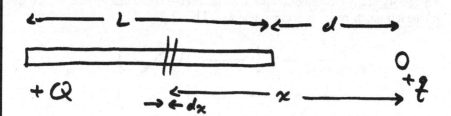

We need to know the distance from each piece to the charge q. We'll use x for this distance, and x will go from a minimum of d to a maximum of $d + L$. The width of each piece is then dx, the distance from one piece of the rod to the next. Choosing a way to divide the object into pieces and describe each piece with a variable is usually the hardest part of any integral. The force each piece of the rod creates on q is

$$F = \frac{k \text{ (charge of the piece) } q}{(\text{distance to the piece})^2}$$

We need to know the charge in each piece. If the size of the piece dx is $1/100$ of L then the charge is $1/100$ of Q. The size of each piece is dx/L of the length so the charge is dx/L of the charge, or $Q(dx/L)$. Another way to think about this is in terms of density. The density of charge along the rod is $\lambda = Q/L$. The charge in a piece of length dx is $(Q/L)\,\mathrm{d}x$.

$$F = \int_{d}^{d+L} \frac{k\,(Q\,\mathrm{d}x/L)\,q}{(x)^2} = \frac{kQq}{L} \int_{d}^{d+L} \frac{\mathrm{d}x}{(x)^2}$$

$$F = \frac{kQq}{L} \left[-\frac{1}{x} \right]_{d}^{d+L}$$

$$F = \frac{kQq}{L} \left[\left(-\frac{1}{d+L} \right) - \left(-\frac{1}{d} \right) \right]$$

$$F = \frac{kQq}{L} \left(\frac{L}{d(d+L)} \right) = \frac{kQq}{d(d+L)}$$

Does this have the correct units? Coulombs times coulombs times k, divided by meters times meters, equals newtons.

8.2 ELECTRIC FIELDS

Electric forces are closely related to electric fields. A field, in mathematics, is a vector at each point in space. When we speak of the altitude of the land, we mean that at any point in the two-dimensional (north–south, east–west) landscape the terrain has a height, typically measured compared to sea level. Imagine that instead, at each spot on the terrain, there was a vector—that would be a field.

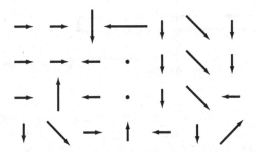

In the last section we said that charges create forces on other charges. That was a simplification. Charges create electric field, and electric field creates forces on charges. Before we needed a way to find the direction and magnitude of the electric force. Now we need a way to find the direction and magnitude of the electric field, and the direction and magnitude of the force that the electric field creates on a charge.

We start with the second part. The electric force on a charge is

$$F = qE$$

where E is the electric field at the spot where the charge is. If q is positive, then E and F point in the same direction. If q is negative, then E and F point in opposite directions.

Now we can figure out the first part. If we have two positive charges, they repel one another—the force on the second is away from the first. But the force on the second is in the same direction as the field there, so the field at the second charge is away from the first charge. Repeat this with a negative and a positive charge, and the field is toward the negative charge. The field is

$$E = \frac{F}{q} = \frac{kQ_1Q_2/R^2}{Q_2} = \frac{kQ}{R^2}$$

There are four ways that one might find an electric field:

- Find the field from each individual charge and add the fields (as vectors).
- Take the force on a charge and divide by the charge to find the field.
- If there is a lot of symmetry, use Gauss's law (next chapter).
- Find the electric field from the electric potential (chapter after next).

EXAMPLE

An electron in an electric field experiences an acceleration of 1.2×10^{16} m/s² northward. What is the electric field that causes this acceleration?

Newton's second law tells us that $F = ma$, but the only force is from an electric field.

$$F = ma$$
$$qE = ma$$
$$E = \frac{m}{q}a = \frac{m}{(-e)}a$$

$$E = \frac{(9.11 \times 10^{-31} \text{ kg})}{(-1.6 \times 10^{-19} \text{ C})}(1.2 \times 10^{16} \text{ m/s}^2 \, \widehat{\text{north}})$$

where $\widehat{\text{north}}$ is a unit vector pointing north, a vector with a magnitude of one and no units.

$$E = (6.8 \times 10^4 \text{ N/C})(-\widehat{\text{north}})$$
$$E = 6.8 \times 10^4 \text{ N/C} \, \widehat{\text{south}}$$

Because the charge experiencing the field is negative, the field that creates the force and the force are in opposite directions.

The units of electric field are newtons per coulomb (N/C). A charge in a 12 N/C field experiences 12 newtons of force for each coulomb of charge it has.

EXAMPLE

What is the electric field halfway between two $+2$ μC charges 20 cm apart?

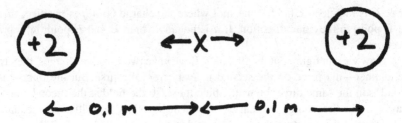

Positive charges create electric field that goes away from the positive charge. So the left charge creates an electric field away from it, or to the right. The right charge creates an electric field away from it, or to the left. The total electric field is

$$E = +\frac{k\,(2\,\mu c)}{(0.1\text{ m})^2} - \frac{k\,(2\,\mu c)}{(0.1\text{ m})^2} = 0$$

The fields have the same magnitude and are in opposite directions, so they add to zero.

LESSON

Anytime we're adding vectors, use the diagram to get the direction and the formula to get the magnitude, just as we did with forces.

EXAMPLE

What is the electric field halfway between a $+2\ \mu C$ charge and a $-2\ \mu C$ charge 20 cm apart?

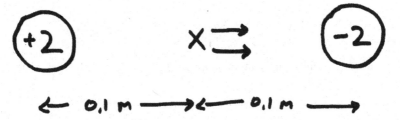

Positive charges create electric field that goes away from the positive charge, so the left charge creates an electric field away from it, or to the right. Negative charges create electric field that goes toward the negative charge, so the right charge creates an electric field toward it, or to the right. The total electric field is

$$E = +\frac{kq}{R^2} + \frac{kq}{R^2}$$

$$E = \frac{(9 \times 10^9 \text{ N m}^2/\text{C}^2)(2 \times 10^{-6}\text{ C})}{(0.1\text{ m})^2} + \frac{(9 \times 10^9 \text{ N m}^2/\text{C}^2)(2 \times 10^{-6}\text{ C})}{(0.1\text{ m})^2}$$

$$E = +3.6 \times 10^6 \text{ N/C}$$

where the positive direction is toward the negative charge. Why did we use $+2 \times 10^{-6}$ C for both the positive charge and the negative charge? The formula kq/R^2 tells us only the magnitude of the electric field. We use the drawing to get the direction, like we always have with free body diagrams.

LESSON

Electric field goes away from positive charges and toward negative charges.

Since electric field and electric force are so similar, and the calculations are nearly the same, why do we bother with electric field at all? Electric fields can do things between when they are created and when they act on a charge. Also, there are some types of calculations that are only possible with electric fields. More on those later.

EXAMPLE

A $+5\ \mu$C charge and a $-2\ \mu$C charge are 30 cm apart. Where is the electric field equal to zero (other than at an infinite distance away)?

Where could the electric field be zero? Anywhere other than on the line joining the charges the fields created by the two charges won't be colinear and they won't add to zero. Between the two charges the electric fields will both be to the right, as in the last example. The only place where the electric field could be zero is on the line joining the charges but outside the charges.

To the left of the $+5\ \mu$C charge, the $+5\ \mu$C charge is both bigger and closer, so it will create a stronger electric field than the $-2\ \mu$C charge will, so the fields won't add to zero. To the right of the $-2\ \mu$C charge, the $+5\ \mu$C charge is bigger but the $-2\ \mu$C charge is closer, so it's possible that the fields have the same magnitude and add to zero.

$$E = +\frac{k(5\ \mu\text{C})}{(30\ \text{cm} + x)^2} - \frac{k(2\ \mu\text{C})}{(x)^2} = 0$$

$$\frac{5}{(30\ \text{cm} + x)^2} = \frac{2}{(x)^2}$$

$$5(x)^2 = 2(30\ \text{cm} + x)^2$$

$$\sqrt{5}(x) = \sqrt{2}(30 \text{ cm} + x)$$

$$(\sqrt{5} - \sqrt{2})x = \sqrt{2}(30 \text{ cm})$$

$$x = \left(\frac{\sqrt{2}}{\sqrt{5} - \sqrt{2}} \right) (30 \text{ cm}) = 52 \text{ cm}$$

There is a spot 52 cm to the right of the $-2 \ \mu C$ charge where the electric field is zero.

EXAMPLE

Two charges of $+2 \ \mu C$ and one of $-2 \ \mu C$ are at three corners of a square. Each side of the square has length 0.3 m. Find the electric field at the fourth corner of the square.

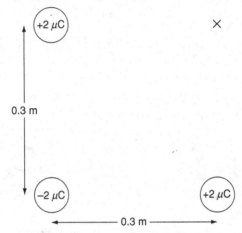

The top left and bottom right charges create electric fields away from themselves, or right and up, respectively. The bottom left charge creates electric field towards itself, or down and to the left. We want to add these three electric field vectors, so we need to choose a pair of axes. We could use the x- and y-axes, or the a- and b-axes.

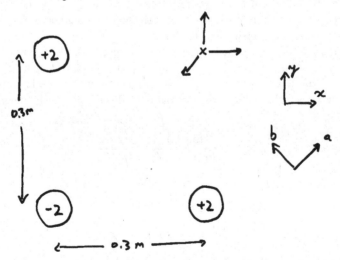

First, using the x- and y-axes (remember that $\cos 45° = 1/\sqrt{2}$):

$$E_x = +\frac{kQ}{L^2} - \frac{kQ}{(\sqrt{2}L)^2}\cos 45° = \frac{kQ}{L^2}\left(1 - \frac{1}{2\sqrt{2}}\right)$$

$$E_y = -\frac{kQ}{(\sqrt{2}L)^2}\cos 45° + \frac{kQ}{L^2} = \frac{kQ}{L^2}\left(1 - \frac{1}{2\sqrt{2}}\right)$$

where $Q = 2\,\mu C$ and $L = 0.3$ m is the length of a side of the square. The total electric field is

$$E = \sqrt{E_x^2 + E_y^2}$$

$$E = \sqrt{\left[\frac{kQ}{L^2}\left(1 - \frac{1}{2\sqrt{2}}\right)\right]^2 + \left[\frac{kQ}{L^2}\left(1 - \frac{1}{2\sqrt{2}}\right)\right]^2}$$

$$E = \sqrt{2}\left(\frac{kQ}{L^2}\right)\left(1 - \frac{1}{2\sqrt{2}}\right)$$

$$E = \left(\frac{(9 \times 10^9 \text{ N m}^2/C^2)(2 \times 10^{-6} \text{ C})}{(0.3 \text{ m})^2}\right)\left(\sqrt{2} - \frac{1}{2}\right)$$

$$E = 1.8 \times 10^5 \text{ N/C}$$

Using the a- and b-axes:

$$E_a = +\frac{kQ}{L^2}\cos 45° + \frac{kQ}{L^2}\cos 45° - \frac{kQ}{(\sqrt{2}L)^2}$$

$$E_b = -\frac{kQ}{L^2}\cos 45° + \frac{kQ}{L^2}\cos 45° = 0$$

$$E = \sqrt{E_a^2 + E_b^2} = E_a = \frac{kQ}{L^2}\left(\frac{1}{\sqrt{2}} + \frac{1}{\sqrt{2}} - \frac{1}{2}\right)$$

$$E = \left(\frac{(9 \times 10^9 \text{ N m}^2/C^2)(2 \times 10^{-6} \text{ C})}{(0.3 \text{ m})^2}\right)\left(\sqrt{2} - \frac{1}{2}\right)$$

$$E = 1.8 \times 10^5 \text{ N/C}$$

The problem has symmetry—draw a diagonal line from the negative charge to the fourth corner, and if you flip the problem it looks the same, so the answer looks the same, so the electric field must be on that diagonal line.

LESSON

Use the same vector-type techniques to find the field as you would to find the force.

EXAMPLE

A $+1\,\mu C$ charge is 1.30 m over a second charge q. A third charge of $-1\,\mu C$ floats midway between the first two charges. The mass of the negative charge is $m = 0.004$ kg. What is the value of q?

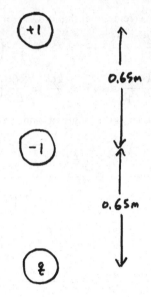

Taking upward as the positive direction, the force on the negative charge is

$$F = ma$$

$$+ \frac{k(1\ \mu C)(1\ \mu C)}{(0.65\ m)^2} - \frac{kq(1\ \mu C)}{(0.65\ m)^2} - mg = 0$$

The sign of the second term is chosen because if q is positive the force will be in the negative direction. If q is negative, the electric force would be upward, but the term in the equation will then be positive. So we can solve this for q and the sign of q will work out.

$$\frac{k(1\ \mu C - q)(1\ \mu C)}{(0.65\ m)^2} = mg$$

$$(1\ \mu C - q) = \frac{mg(0.65\ m)^2}{k(1\ \mu C)}$$

$$(1\ \mu C - q) = \frac{(0.004\ kg)(9.8\ m/s^2)(0.65\ m)^2}{(9 \times 10^9\ N\ m^2/C^2)(1\ \mu C)}$$

$$(1\ \mu C - q) = 1.84\ \mu C$$

$$q = -0.84\ \mu C$$

The electric force from the top charge is not enough to hold the middle charge up, so the electric force from the bottom charge is upward, or repulsive.

Electric forces on small charges are often much bigger than the force of gravity. Therefore we usually ignore gravity in electric problems. If a problem mentions which direction is up, then gravity is probably a significant force in the problem; otherwise, it probably isn't.

EXAMPLE ⋆

A charge of $+Q$ is spread out uniformly over a rod of length L. The point P is a distance d above the midpoint of the rod. What is the electric field at point P?

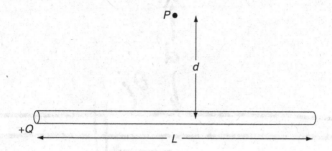

We would like to use $E = kq/R^2$, but not all of the charge is the same distance from point P. We could divide the rod in half, find the electric field from each half, and add them (as vectors). We could divide the rod into three pieces for a better answer, or four or more. We want to divide the rod into pieces so small that we can say that all of the charge in each piece is the same distance from the point P. We'll divide the rod into an infinite number of infinitely small pieces, find the electric field from each piece, and add the electric field vectors.

The magnitude of the electric field that each piece of the rod creates at P is

$$E = \frac{k \text{ (charge of the piece)}}{(\text{distance to } P)^2}$$

We need to know the distance from each piece to the point P and the charge of each piece.

We'll use x for the distance from the midpoint of the rod to a piece of the rod. Thus, x will go from a minimum of $-L/2$ to a maximum of $+L/2$. The width of each piece is then dx, the distance from one piece of the rod to the next. The distance from each piece to P is

$$R = \sqrt{d^2 + x^2}$$

We need to know the charge in each piece. If the size of the piece dx is $1/100$ of L then the charge is $1/100$ of Q. The size of each piece is dx/L of the length so the charge is dx/L of the charge, or $Q(dx/L)$. Alternatively, the density of charge along the rod is $\lambda = Q/L$ and the charge in a piece of length dx is $(Q/L)\, dx$.

The vertical component of the electric field at P is

$$E_y = \int_{-L/2}^{+L/2} \frac{k\,(Q\,dx/L)}{(\sqrt{d^2 + x^2})^2} \sin\theta$$

where θ is the angle in the figure. (The horizontal component is zero because of the symmetry of the problem.) It appears that we need to find the angle θ for each piece

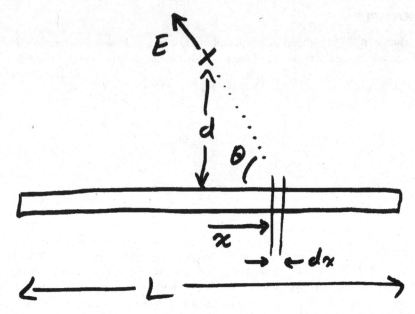

of the rod. But we don't need θ, we need $\sin\theta$. We have a right triangle in the figure that includes the angle θ, so

$$\sin\theta = \frac{\text{opposite}}{\text{hypotenuse}} = \frac{d}{\sqrt{d^2+x^2}}$$

$$E_y = \int_{-L/2}^{+L/2} \frac{k\,(Q\,dx/L)}{(\sqrt{d^2+x^2})^2}\,\frac{d}{\sqrt{d^2+x^2}} = \frac{kQd}{L}\int_{-L/2}^{+L/2}\frac{dx}{(\sqrt{d^2+x^2})^3}$$

$$E_y = \frac{kQd}{L}\left[\frac{x}{d^2\sqrt{d^2+x^2}}\right]_{-L/2}^{+L/2}$$

$$E_y = \frac{kQd}{L}\left[\left(\frac{(L/2)}{d^2\sqrt{d^2+(L/2)^2}}\right) - \left(\frac{-(L/2)}{d^2\sqrt{d^2+(-L/2)^2}}\right)\right]$$

$$E_y = \frac{kQ}{d\sqrt{d^2+(L/2)^2}}$$

8.3 MAGNETIC FORCES

Charges create electric field, and electric field creates forces on charges. Moving charges create magnetic field, and magnetic field creates forces on moving charges. When lots of charges move in the same direction, we call it current. Current is measured in amperes,

$$1\,\text{A} = 1\,\text{C/s}$$

and magnetic field is measured in teslas,

$$1\ T = 1\ N/A\ m = 1\ N\ s/C\ m$$

Since current is moving charges, current creates magnetic field. We need a way to find the direction and magnitude of the magnetic field, and the direction and magnitude of the force that the magnetic field creates.

We again start with the second part. The magnetic force on a moving charge is always perpendicular to the magnetic field and is always perpendicular to the velocity of the charged particle. Imagine a proton moving out of the page in a magnetic field that points left to right. Is there any direction that is perpendicular to both out of the page and right at the same time? Only up the page and down the page. Note that any magnetic field problem must be three-dimensional.

To determine whether the magnetic force is into the page or out of the page. Hold your hand so that the fingers point out of the page (as in the figure). Keep your fingers pointing out of the page and rotate your hand so that the palm points right—curl your fingers from out to right. Where does your thumb point? Up the page.

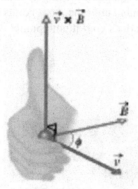

Drawing a vector up the page or right is easy, but drawing one into or out of the page is less so. To describe a vector out of the page we use the dot of the arrow coming at us (⊙ symbol), like in a 3-D movie. To describe a vector into the page we use the tail feathers of the arrow going away from us (⊗ or × symbol).

The magnetic force is

$$F = qv \times B$$

where B is the magnetic field at the spot where the charge is. The × symbol means "cross-product;" the cross-product of two vectors is a third vector that is perpendicular to both of the original two vectors. We use the "right-hand rule" to find the direction, as above. The magnitude of the cross-product is the magnitude of the first vector times the magnitude of the second vector times the sine of the angle between the vectors. So, often we'll use

$$F = qvB \sin \theta$$

If q is negative then that reverses the direction of the magnetic force. Also, if v and B are parallel, then the angle between them is zero and there is no magnetic force.

EXAMPLE

An electron travels across the paper left to right. There is a magnetic field going up the page (which is not the same as north). What is the direction of the magnetic force on the electron?

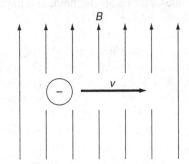

Take your right hand and point the fingers to the right. Keeping your fingers in that direction, rotate your hand until the palm points up. Now curl your fingers from right to up. What direction does your thumb point? It should point out of the page.

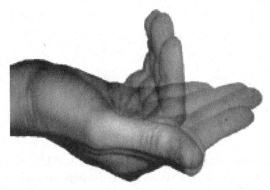

But the electron has a negative charge, and that reverses the direction. The magnetic force is into the page.

EXAMPLE

An electron travels horizontally to the south at 2×10^5 m/s. There is a 0.04 T magnetic field going northward and 40° below horizontal. What is the magnetic force on the electron?

Take your right hand and point the fingers to the right (south for this problem). Keeping your fingers in that direction, rotate your hand until the palm points down. Now curl your fingers from the velocity direction to the magnetic field direction (never reverse the order). What direction does your thumb point? It should point into the page. But the electron has a negative charge, and that reverses the direction, so the magnetic force is out of the page, which is westward.

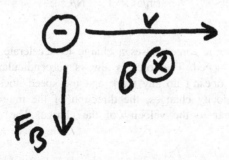

The magnitude of the magnetic force is

$$F = qvB \sin \theta$$
$$F = (1.6 \times 10^{-19} \text{ C})(2 \times 10^5 \text{ m/s})(0.04 \text{ T}) \sin 140° = 8.2 \times 10^{-16} \text{ N}$$

We could also have used 40° and gotten the same result.

LESSON

Use the right-hand rule to find the direction of the magnetic force and the formula to find the magnitude, as always.

EXAMPLE

An electron travels horizontally to the right at 2×10^5 m/s. There is a 0.04 T magnetic field pointing into the page. What electric field could be applied to the electron so that its acceleration would be zero?

Take your right hand and point the fingers to the right. Keeping your fingers in that direction, rotate your hand until the palm points into the page. Now curl your fingers from the velocity direction to the magnetic field direction. What direction does your thumb point? It should point up the page (which is not the same as north). But the electron has a negative charge, and that reverses the direction, so the magnetic force is down the page.

Adding the forces on the electron

$$F = ma$$

$$F_B + F_E = 0$$

$$evB \sin 90° \ \overset{\frown}{\textbf{down}} + F_E = 0$$

Notice that we have replaced the charge q with e. We have already accounted for the sign of the electron's charge, so we want only the magnitude in the equation.

The electric force is

$$F_E = qE$$

$$qE = -evB \sin 90° \ \overset{\frown}{\textbf{down}}$$

$$qE = evB \ \overset{\frown}{\textbf{up}}$$

$$(-e)E = evB \ \overset{\frown}{\textbf{up}}$$

$$E = vB \ \overset{\frown}{\textbf{down}}$$

The magnetic force on the electron is down, so the electric force must be up so that the forces add to zero. To get an electric force on a negative charge to go up, the electric field must be down.

$$E = (2 \times 10^5 \text{ m/s})(0.04 \text{ T}) \ \overset{\frown}{\textbf{down}} = 8000 \text{ N/C} \ \overset{\frown}{\textbf{down}}$$

Not sure that m/s times T is N/C? When I can't remember what a unit is equal to (and I usually can't for teslas), I start with an equation I know is good and replace the variables with the units.

$$F = qvB$$

$$\text{N} = (C)(m/s)(T)$$

$$\text{T} = \text{N s/C m}$$

$$(m/s) \times \text{T} = \text{N/C}$$

When a magnetic force causes a charge to accelerate, it changes the direction but not the speed. The force is always perpendicular to the motion, so the magnetic force doesn't do any work and the speed doesn't change. As the direction of the velocity changes, the direction of the magnetic force changes to stay perpendicular to the velocity of the charge. The charge curves in a circle.

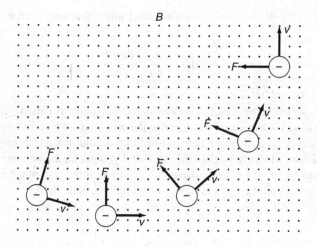

EXAMPLE

An electron travels perpendicular to a 0.04 T magnetic field at 2×10^5 m/s. What is the radius of the circular path of the electron?

The only force is from the magnetic field, so

$$F = ma$$

$$qvB \sin 90° = m\frac{v^2}{R}$$

$$R = \frac{mv}{qB}$$

$$R = \frac{(9.1 \times 10^{-31} \text{ kg})(2 \times 10^5 \text{ m/s})}{(1.6 \times 10^{-19} \text{ C})(0.04 \text{ T})}$$

$$R = 2.8 \times 10^{-5} \text{ m} = 28 \ \mu\text{m}$$

LESSON

Treat magnetic forces as you would any other forces: draw the free body diagram, add the force vectors, and use $F = ma$.

8.4 MAGNETIC FIELDS

Moving charges, or currents, create magnetic fields. A single moving charge creates only a tiny magnetic field, so we'll only worry about the fields created by currents. We need a way to determine the direction and magnitude of the magnetic field created by a current. We do this with the Biot–Savart law (they were French, so the t's are silent).

$$dB = \frac{\mu_0}{4\pi} \frac{i \, ds \times r}{r^3}$$

Here $\mu_0 = 4\pi \times 10^{-7}$ T m/A.

Let's try to explain this. Divide the current i up into tiny pieces of length ds. The vector r is from the tiny piece of current to the spot where we're finding the magnetic field. The cross-product d$s \times r$ is found just like we did in the last section, using the right-hand rule. Multiply by the other stuff and you get the tiny amount of magnetic field created by that tiny piece of current. Add up all the tiny magnetic fields (and there're a lot of such tiny pieces) and you get the magnetic field.

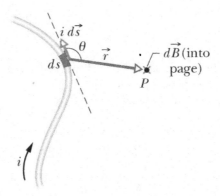

Finding electric fields wasn't this complicated, at least at the beginning. The difference is that each piece of the current is usually a different distance from the spot of interest, and so we have to use calculus to do this.

EXAMPLE ⋆

An infinitely long wire carries a current of i. What is the magnetic field this current creates at the point P a distance d away?

Each piece of the current is a different distance from the point P. We need to divide the wire into pieces so small that we can say that all of each piece is the same distance from the point P. We'll divide the wire into an infinite number of infinitely small pieces, find the magnetic field from each piece and add the magnetic field vectors.

$$B = \int \frac{\mu_0}{4\pi} \frac{i \, ds \times r}{r^3}$$

For each piece, the vector ds points to the right and the vector r points up and either left or right.

$$\overrightarrow{\text{right}} \times (\overrightarrow{\text{up}} + \overrightarrow{\text{right}}) = (\overrightarrow{\text{right}} \times \overrightarrow{\text{up}}) + (\overrightarrow{\text{right}} \times \overrightarrow{\text{right}}) = \overrightarrow{\text{out}} + 0 = \overrightarrow{\text{out}}$$

Each piece of the current creates a magnetic field that points out of the page. Since all of the tiny magnetic fields are in the same direction, they are easy to add and the magnitude of the magnetic field is

$$B = \int dB = \frac{\mu_0 i}{4\pi} \int \frac{ds \; r \sin \theta}{r^3}$$

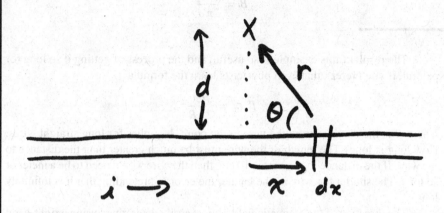

We'll use x for the position of each tiny piece of current, measuring from the midpoint of the wire (the spot closest to the point P). Thus, x will go from $-\infty$ to $+\infty$. The length of each piece is then dx, the distance from one piece of the wire to the next. The distance from each piece to P is

$$R = \sqrt{d^2 + x^2}$$

$$B = \frac{\mu_0 i}{4\pi} \int_{-\infty}^{+\infty} \frac{\sin \theta \; dx}{(\sqrt{d^2 + x^2})^2}$$

We don't need θ, but $\sin \theta$. We have a right triangle in the figure that includes the angle θ, so

$$\sin \theta = \frac{\text{opposite}}{\text{hypotenuse}} = \frac{d}{\sqrt{d^2 + x^2}}$$

$$B = \frac{\mu_0 i d}{4\pi} \int_{-\infty}^{+\infty} \frac{dx}{(\sqrt{d^2 + x^2})^3}$$

$$B = \frac{\mu_0 i d}{4\pi} \left[\frac{x}{d^2 \sqrt{d^2 + x^2}} \right]_{-\infty}^{+\infty}$$

$$B = \frac{\mu_0 i d}{4\pi} \left[\left(\frac{(+\infty)}{d^2 \sqrt{d^2 + (+\infty)^2}} \right) - \left(\frac{(-\infty)}{d^2 \sqrt{d^2 + (-\infty)^2}} \right) \right]$$

Infinity plus d^2 is the same as infinity. The square root of $(\infty)^2$ is ∞, and ∞ divided by ∞ is one.

$$B = \frac{\mu_0 i d}{4\pi}\left[\left(\frac{1}{d^2}\right) + \left(\frac{1}{d^2}\right)\right]$$

$$B = \frac{\mu_0 i}{2\pi d}$$

The result of this example is so useful, and the process of getting it so long (or painful, if you prefer), that even physicists learn the formula

$$B_{\text{wire}} = \frac{\mu_0 i}{2\pi R}$$

where R is the distance from the wire to the point. It applies for long straight wires. How long is long? The length of the wire must be much greater than the distance to the wire. If the distance to the wire is 10 cm, then the wire would need to be a meter or so long. The shorter the wire is, the greater the error in pretending that it is infinitely long.

The direction of the magnetic field is most easily determined using a right-hand rule (not the same right-hand rule as before). The magnetic field goes in loops around the current—electric field lines start and end at charges but magnetic field lines never end. Which way around do the magnetic field lines go? Take your right hand and point your thumb in the direction of the current, to the right in the last example. Curl your fingers—which way do they go? Out of the page above the wire, down in front of the wire, into the page below the wire, and up behind the wire.

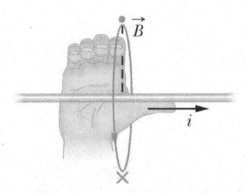

It is not strange that the magnetic field direction is not a simple "up" or "left." The direction of the electric field was "away from the charge" or "toward the charge," which could mean up or down depending on where we were and where the charge was. The direction of the magnetic field depends on where we are and where the current is.

EXAMPLE

What is the magnetic field halfway between two wires 20 cm apart, each of which carries a current of 15 A, if the currents are in (a) the same direction and (b) in opposite directions?

We look at the wires head-on, with both currents out of the page.

What is the direction of the magnetic field created by the left current? Take your right hand and hold it with the thumb (current) pointing toward you. Curl your fingers—they go around counterclockwise. The magnetic field goes around the current counterclockwise. At the spot midway between the wires this magnetic field is up the page. Do the same for the right current and it will be down the page.

 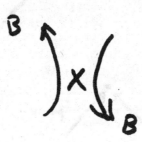

Now we add the two magnetic field vectors. Since each wire has the same current and is the same distance from the spot, the magnitude of each magnetic field is the same. They point in opposite directions, so one is in the positive direction and one is in the negative direction. They add to zero, so there is no magnetic field at the midpoint between the wires.

What if the right-hand wire carried current into the page? Point your right thumb away from you and curl your fingers. They go clockwise, so the magnetic field goes clockwise around the current, which is up at the spot of interest.

 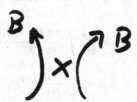

The magnetic field is

$$B = B_{\text{left}} + B_{\text{right}}$$

$$B = \frac{\mu_0 i}{2\pi R}\,\widehat{\text{up}} + \frac{\mu_0 i}{2\pi R}\,\widehat{\text{up}} = \frac{2\mu_0 i}{2\pi R}\,\widehat{\text{up}}$$

$$B = \frac{(4\pi \times 10^{-7}\,\text{T m/A})(15\,\text{A})}{\pi(0.10\,\text{m})}\,\widehat{\text{up}}$$

$$B = 6 \times 10^{-5}\,\text{T}\,\widehat{\text{up}} = 60\,\mu\text{T}\,\widehat{\text{up}}$$

This is about the same magnitude as Earth's magnetic field.

EXAMPLE ★

A current i goes counterclockwise around a loop of radius R. What is the magnetic field at the center of the loop?

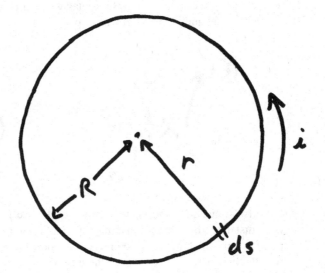

Each piece of the current is the same distance from the center of the circle. The direction of the magnetic field created by each piece is out of the page. Also, ds is perpendicular to the vector r for each tiny piece of current.

$$B = \int \frac{\mu_0}{4\pi}\,\frac{i\,\text{d}s \times r}{r^3}$$

$$B = \frac{\mu_0 i}{4\pi} \int \frac{\text{d}s\,R\,\sin 90^\circ}{R^3}$$

$$B = \frac{\mu_0 i}{4\pi R^2} \int \text{d}s$$

The variable ds is the length of each tiny piece, and $\int ds$ is the sum of the lengths of each tiny piece. This is the distance around the loop, the circumference $2\pi R$.

$$B = \frac{\mu_0 i}{4\pi R^2} \, 2\pi R$$

$$B = \frac{\mu_0 i}{2R}$$

While this result is not as difficult to get as the last one, it is another formula that's often remembered.

$$B_{\text{loop}} = \frac{\mu_0 i}{2R}$$

To get the direction of the magnetic field, pick a spot on the wire loop and use the right-hand rule: thumb in the direction of the current and curl your fingers. Alternatively, curl your fingers with the current (counterclockwise) and your thumb will point in the direction of the magnetic field in the center of the loop (out of the page). Either method works, as long as you don't use your left hand (which is easy to do if your pencil is in your right hand).

EXAMPLE

What is the magnetic field in the center of the arc? The arc has a radius of 6 cm and the wire carries 5 A of current.

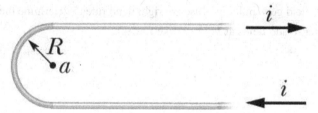

We divide the current into three parts, find the magnetic field from each part, and add the fields.

$$B = B_{\text{top}} + B_{\text{arc}} + B_{\text{bottom}}$$

The top is half of an infinitely long straight wire. Pointing the thumb in the direction of the current (right), the fingers curl into the page below the wire. The bottom is also half of an infinitely long straight wire. Pointing the thumb in the direction of the current (left), the fingers curl into the page above the wire. The arc is half of a loop, also creating a field into the page.

$$B = \left(\frac{1}{2} \frac{\mu_0 i}{2\pi R} + \frac{1}{2} \frac{\mu_0 i}{2R} + \frac{1}{2} \frac{\mu_0 i}{2\pi R} \right) \, \widehat{\text{into}}$$

$$B = \frac{\mu_0 i}{4\pi R} (1 + \pi + 1) \, \widehat{\text{into}}$$

$$B = \frac{(4\pi \times 10^{-7} \text{ T m/A})(5 \text{ A})}{4\pi (0.06 \text{ m})} (2 + \pi) \; \widehat{\text{into}}$$

$$B = 4.3 \times 10^{-5} \text{ T } \widehat{\text{into}} = 43 \; \mu\text{T } \widehat{\text{into}}$$

LESSON

Finding magnetic fields using the Biot–Savart law is difficult. Instead try to use the results for B_{wire} and B_{loop}.

CHAPTER SUMMARY

- Charge creates electric field, and electric field creates forces on charges.

$$Q \xrightarrow{E=KQ/R^2} E \xrightarrow{F=qE} F_{\text{electric}}$$

- Moving charge creates magnetic field, and magnetic field creates forces on moving charges.

$$I \xrightarrow{B=\mu_0 I/2\pi R} B \xrightarrow{F=qvB\cos\theta} F_{\text{magnetic}}$$

- Electric field goes from positive to negative charge.

- Magnetic field goes in loops—use the right hand rule to determine the direction.

- Remember to add fields as vectors (like forces).

CHAPTER 9

ADVANCED ELECTRIC AND MAGNETIC FIELDS

In the last chapter, we found the electric (and magnetic) fields by adding the individual electric fields created by each charge. If there were a lot of charges and they were not all the same distance from our point, this could get messy. Gauss's law can be used to simplify some of these cases.

Sometimes it is helpful to visualize electric or magnetic fields with a series of lines. To do this, find the electric field at one spot, go a little in that direction, find the electric field at the new spot, and continue doing this until the line ends at a negative charge. For a constant electric field, you get a series of straight lines. If the field is due to a single positive charge, you get a series of lines going away from the charge. The greater the density of "field lines" the stronger the field, so for a point charge (all of the charge at a single point), as you go farther away and the spacing between lines increases, the field strength decreases.

The figure shows the electric field lines from a pair of charges, one positive and the other negative. Field lines come out of the positive charge, go away from the positive charge, go toward the negative charge, and end at the negative charge. The number of field lines starting at the positive charge is the same as the number that end at the negative charge because they have the same amount of charge.

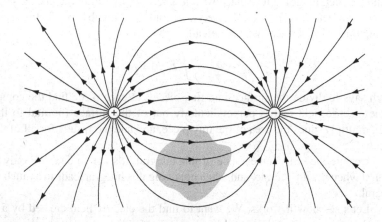

Now consider the shaded area between the two charges. How many electric field lines come out of the shaded area? Remember to count a field line going in as a negative field line going out. The total number of field lines coming out is zero—for every field line that does come out there's one going in. The only way to get more coming out than going in is if there is a source of lines in the area—a positive charge. The more positive charge in the area the more lines created and the more lines coming out. Likewise the only way to get a negative number of lines coming out (more in than out) is if there is a "sink" of lines in the area—a negative charge.

Gauss's law states that the number of electric field lines coming out of any shaded area is equal to the net charge inside the area. The trick is to use the count of field lines to determine the electric field. Similar tricks can be done with Ampere's law for magnetic fields.

9.1 GAUSS'S LAW ⋆

The mathematical expression for Gauss's law is

$$\Phi_E = \frac{q_{net}}{\epsilon_0}$$

where Φ_E is the electric field flux coming out of the area, q_{net} is the net charge inside the area, and ϵ_0 is $1/4\pi k = 8.85 \times 10^{-12}$ C^2/N m^2.

Gauss's law works for any closed surface. A closed surface is one with no exits: if you are inside the surface and you want to get out you need to pass through the surface. The surface does not have to be real—imagine or invent any surface you want and Gauss's law tells you the number of electric field lines coming out.

What is flux? Imagine that you are out in the rain holding a bucket. Flux would be the rate at which raindrops enter the bucket. You could reduce this rate by using a smaller bucket, by going to a place with less rain, or by tilting the bucket sideways. The electric field flux Φ_E is effectively a count of the number of electric field lines passing through a surface. Mathematically,

$$\Phi_E = \int E \cdot dA$$

which says to divide the surface into tiny pieces, at each piece find the component of the electric field that is perpendicularly out of the surface (normal to the surface), multiply times the area of the tiny piece, and add the results for all of the tiny pieces.

Despite the scary appearance of this integral, it's not so bad. We only do the integral when it's dirt simple, and when not doing this integral leads to a much nastier integral.

Let's see how it works. We want to find the electric field created by a single point charge $q = +2$ μC. What is the electric field a distance r away? We can invent any surface we want to use with Gauss's law, so we'll use a sphere, centered on the charge q with a radius of r (r is a variable, it could be anything).

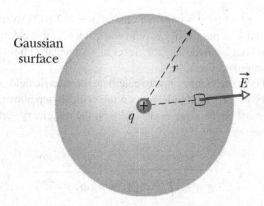

Gaussian
surface

We want to do the integral to find the flux. To do that we need to know the electric field at each spot on our imaginary surface. We don't know the field—that's what we're trying to find. What we do know is that the field on one side of the imaginary sphere is the same as the field on the other side. Pick any two spots on the imaginary sphere, and the electric field is the same. It has to be because of the symmetry of the problem—it looks the same from any angle, so the electric field is the same at any angle.

At any spot on our imaginary sphere the electric field is perpendicularly out of the surface and has the same magnitude. Because the field is perpendicularly out

$$E \cdot dA = E \, dA \, \cos 0° = E \, dA$$

Because the field E is the same everywhere on our imaginary "Gaussian" surface

$$\Phi_E = \int E \, dA = E \int dA$$

The integral that remains is to add up the area of each tiny piece, so it's equal to the area of the imaginary surface.

$$\Phi_E = E \int dA = E A$$

Gauss's law tells us that the flux is equal to the charge "enclosed" by the imaginary surface divided by ϵ_0.

$$E A = \Phi_E = \frac{q_{net}}{\epsilon_0}$$

Because the surface is a sphere, the area is $4\pi r^2$.

$$E(4\pi r^2) = \frac{q_{net}}{\epsilon_0}$$

$$E = \frac{q_{net}}{4\pi r^2 \epsilon_0} = \frac{q}{(4\pi \epsilon_0)r^2} = \frac{k(+2 \, \mu C)}{r^2}$$

This is the same thing we got in the last chapter (whew).

What if we had used a negative charge $q = -2 \ \mu$C? Everything would have been the same until we put the negative in and got a negative electric field. We always measure the field out from the surface, never in, so a negative field is inward or toward the negative charge.

The goal is to use Gauss's law to calculate the electric field. We can only do that if the symmetry in the problem lets us do the critical step above. The electric field, whatever it is, must be the same everywhere on the imaginary surface so we can pull it out of the integral.

$$\overbrace{EA = \underbrace{\int E \cdot dA}}^{\text{symmetry}} \; = \; \overbrace{\Phi_E = \frac{q_{\text{net}}}{\epsilon_0}}^{\text{Gauss's law}}$$

$$\underbrace{}_{\text{definition of flux}}$$

EXAMPLE

What is the electric field a distance r away from a charge Q spread uniformly over a sphere of radius R?

Being able to use Gauss's law depends on finding a "good" imaginary surface. We need one where the electric field is the same everywhere on the surface, even if we don't know what that electric field is. The charge has spherical symmetry—we can rotate it any way we want about the center point and it looks the same. Therefore we choose as our surface a sphere of radius r centered on the sphere of charge.

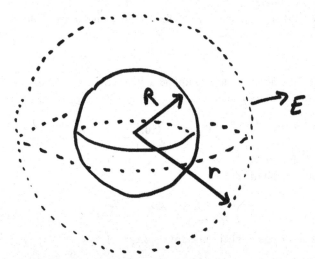

Now we have the power of symmetry to help us do the integral.

$$EA = \Phi_E = \frac{q_{\text{net}}}{\epsilon_0}$$

$$E(4\pi r^2) = \frac{q_{\text{net}}}{\epsilon_0}$$

The question is how much charge is enclosed. If $r \geq R$, then all of the charge is enclosed.

$$E(4\pi r^2) = \frac{Q}{\epsilon_0}$$

$$E = \frac{Q}{4\pi \epsilon_0 r^2} = \frac{kQ}{r^2}$$

As long as the object has spherical symmetry, the electric field is the same as if it were a point charge.

If $r < R$, and we are finding the electric field at some spot inside the ball of charge, then only some of the charge is enclosed. If the volume of the imaginary surface is half the volume of the ball of charge, then it encloses half of the charge (this is because the charge is spread *uniformly*). The charge enclosed by the surface is then

$$q_{net} = \frac{\text{volume of surface}}{\text{volume of ball}} Q = \frac{\frac{4}{3}\pi r^3}{\frac{4}{3}\pi R^3} Q = \frac{r^3}{R^3} Q$$

$$E(4\pi r^2) = \left(\frac{r^3}{R^3} Q\right) \frac{1}{\epsilon_0}$$

$$E = \frac{Qr}{4\pi \epsilon_0 R^3} = \frac{kQr}{R^3}$$

So

$$E(r) = \begin{cases} kQr/R^3, & \text{for } r < R \\ kQ/r^2, & \text{for } r \geq R \end{cases}$$

For comparison, if we had tried to do the last example using the technique from the last chapter, dividing the sphere into an infinite number of tiny pieces, we would have eventually had to solve this integral:

$$E = \frac{3kQ}{2R^3} \int_0^\pi d\theta \int_0^R dt \, \frac{t^2(r - t\sin\theta)\sin\theta}{(r^2 + t^2 - 2rt\sin\theta)^{3/2}}$$

The purpose of Gauss's law is to replace a fiendish integral with a simple one! There are only three geometries where the symmetry lets us make use of Gauss's law.

EXAMPLE

What is the electric field created by a long line of charge with uniform linear density λ (charge per length)?

We can't rotate a line about the center point and have it look the same. We can spin it about its axis and it will be the same no matter how much we spin it. This is called cylindrical symmetry, and we choose a cylinder as our imaginary surface. Our cylinder has a radius of r and a height (or length) of h.

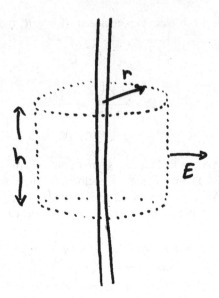

Because the surface must be closed, the cylinder has a top and bottom. The electric field going down through the top from the charge above is equal to the electric field going up through the top from the charge below. So the flux through the top and bottom of the cylinder is zero. The electric field through the side is the same everywhere.

$$EA = \Phi_E = \frac{q_{net}}{\epsilon_0}$$

The charge enclosed by the imaginary surface is the density of charge (charge per length) times the length enclosed.

$$E(2\pi r h) = \frac{\lambda h}{\epsilon_0}$$

$$E = \frac{\lambda}{2\pi \epsilon_0 r} = \frac{2k\lambda}{r}$$

EXAMPLE

What is the electric field created by a large plane of charge with density σ (charge per area)?

With a very large plane of charge, the electric field will go perpendicularly out from the plane. Since all of the electric field lines are parallel, the density of field lines doesn't change, and the electric field magnitude stays the same. Shouldn't the field get smaller as we get further away from the charge? As long as the plane is much bigger than the distance to the plane, the field stays the same.

We choose as our imaginary surface a cylinder (soda can) with the surface cutting the cylinder in two crossways. Because the field goes perpendicularly away from the plane, there is no flux through the curved sides of the "can." The field there has no component perpendicular to the surface—it skims the surface but doesn't go through.

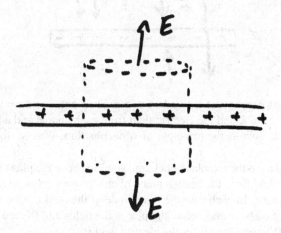

The electric field through the circular ends is the same everywhere on the end. Also, the field at each end is the same.

$$E A = \Phi_E = \frac{q_{\text{net}}}{\epsilon_0}$$

$$E(2a) = \Phi_E = \frac{a\sigma}{\epsilon_0}$$

Here a is the area of each circular end.

$$E = \frac{\sigma}{2\epsilon_0}$$

LESSON

Gauss's law is always true, but to use it to find the electric field we need a symmetric problem.

EXAMPLE

What is the electric field between two large parallel planes of charge? The top plane has charge density σ and the bottom plane has charge density $-\sigma$.

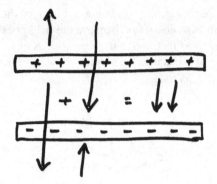

In the last chapter, when we had two point charges and we wanted to know the electric field we found the electric field from each charge and added the electric fields. This is known as the principle of **superposition**. We can do the same thing here.

The field from the top plane of charge goes up above the plane and down below the plane. The field from the bottom plane of charge goes down above the plane and up below the plane. In each case the magnitude of the field is $E = \sigma/2\epsilon_0$ (from the last example). So above and below the planes the fields add to zero, and there is no electric field. Between the planes the electric field is

$$E = E_{\text{top}} + E_{\text{bottom}} = \frac{\sigma}{2\epsilon_0} + \frac{\sigma}{2\epsilon_0} = \frac{\sigma}{\epsilon_0}$$

Whenever the field lines from a plane go in both directions, the field is

$$E_{\text{both}} = \frac{\sigma}{2\epsilon_0}$$

but when the field lines only go in one direction the field is

$$E_{\text{one}} = \frac{\sigma}{\epsilon_0}$$

EXAMPLE

A conducting sphere of radius a has a charge of $+Q$ on it. A point charge $-Q$ is a distance $3a$ away from the center of the sphere. What is the electric field halfway between the point charge and the surface of the sphere?

We've done the electric field due to a sphere, and we've done the electric field due to a point charge. We could do each separately and add the fields (as vectors, of course). That doesn't work here.

To do the sphere we drew an imaginary sphere around the ball of charge. Then we used the symmetry of the problem to say that the electric field was the same everywhere on our imaginary sphere. But the $-Q$ charge exerts a force on the charges in the sphere, pulling positive charge toward it. The charge in the sphere is no longer symmetric. Without the symmetry we can't do the integral to find the flux.

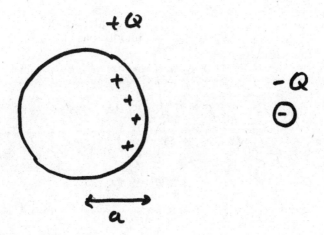

(It is possible to solve this problem, but it requires more advanced mathematics than we are using in this book.)

We could do this if the sphere were made of insulating material instead of conducting material. Then the charges in the sphere could not move around and would be unaffected by the point charge.

LESSON

Superposition lets us find the field of each charge separately and add the fields. If there is a conductor in the problem, superposition may not work. Check to see if one of the charges causes other charges to move and undo the symmetry.

9.2 AMPERE'S LAW ⋆

Just as it is possible to use field lines to envision an electric field, it is possible to use field lines to envision a magnetic field. Magnetic field lines never end, but go in loops. The figure shoes the magnetic field lines from a current that goes into the page. The field lines get further apart as we move away from the current because the field gets weaker.

When we calculated the electric field flux, we got a total that was related to the charge inside the surface. This is because electric field lines start and end at electric charges. Magnetic field lines don't start and end, so the magnetic field flux through any closed surface is always zero. This is good to know, but it doesn't help us find the magnetic field strength.

Instead of using a closed surface, Ampere's law uses a closed loop.

$$\oint B \cdot \mathrm{d}s = \mu_0 I_{\mathrm{enc}}$$

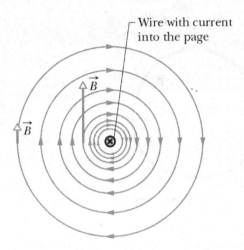

Pick any closed loop—that is, start somewhere and go anywhere so long as you end where you started (using $\oint$ instead of $\int$ is a way to say use a closed path). As you travel along this route, at each point find the magnetic field. Take the component of the magnetic field $\boldsymbol{B}$ that is parallel to your next step ds and multiply them. When you sum the product for all of the steps, Ampere's law says that you get μ_0 times the current enclosed by the loop. This is not the current going along the path, but passing through the surface of which the path is the edge.

Like Gauss's law, we only do the integral when it's dirt simple. Let's see how this works. We want to find the magnetic field created by a long straight wire with current i. We draw an imaginary "Amperian loop" around the wire at a distance r.

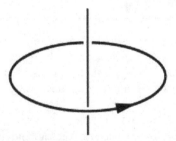

The problem has symmetry—if we rotate the loop along the axis of the wire it looks the same—so the magnetic field is the same everywhere on the imaginary path (if the wire were bent rather than straight then this would not be true). Also, from the right-hand rule this magnetic field will be parallel to the path.

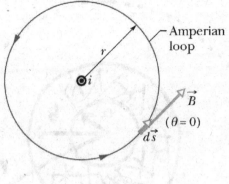

$$\oint \boldsymbol{B} \cdot \mathrm{d}s = \oint B \, \mathrm{d}s = B \oint \mathrm{d}s$$

The last integral $\oint \mathrm{d}s$ is the sum of the lengths of each step for all steps. Because we've already dealt with the vector part (the magnetic field is parallel to the path) it is the length of the steps and not the displacement of the steps. This is the length of the path, so $\oint \mathrm{d}s = L$.

$$\oint \boldsymbol{B} \cdot \mathrm{d}s = \oint B \, \mathrm{d}s = B \oint \mathrm{d}s = BL = B(2\pi r)$$

Ampere's law tells us that this integral is equal to $\mu_0 I_{\mathrm{enc}}$, so

$$B(2\pi r) = \mu_0 I_{\mathrm{enc}} = \mu_0 i$$

$$B_{\mathrm{wire}} = \frac{\mu_0 i}{2\pi r}$$

This is the same result that we got in the last chapter, but this time with simple calculus instead of difficult calculus.

$$\overbrace{BL}^{\text{symmetry}} = \underbrace{\oint \boldsymbol{B} \cdot \mathrm{d}s = \mu_0 I_{\mathrm{enc}}}_{\text{Ampere's law}}$$

EXAMPLE

A long straight hollow tube carries a current of i. The tube has an inner radius of a and an outer radius of b. What is the magnetic field created by the current?

   We choose as our imaginary path a circle or radius r centered on the axis of the tube. Just like with the wire, the magnetic field is parallel to the path and has the same magnitude everywhere on the path. (If the current is out of the page, then the magnetic field is counterclockwise.)

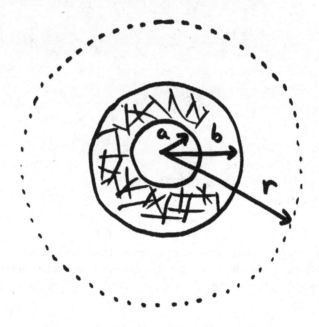

$$BL = \oint \mathbf{B} \cdot \mathrm{d}s = \mu_0 I_{enc}$$

$$B(2\pi r) = \mu_0 I_{enc}$$

How much current is enclosed? If $r < a$, then none of the current is enclosed because the center of the tube is hollow.

$$B(2\pi r) = \mu_0(0)$$

$$B = 0$$

If $r \geq b$, then all of the current is enclosed because the imaginary path is outside the tube.

$$B(2\pi r) = \mu_0(i)$$

$$B = \frac{\mu_0(i)}{2\pi r}$$

If $a \leq r < b$, then some of the current is enclosed.

$$I_{enc} = \frac{(\text{area of tube enclosed})}{(\text{area of tube})} i$$

$$I_{enc} = \frac{\pi r^2 - \pi a^2}{\pi b^2 - \pi a^2} i$$

$$B(2\pi r) = \mu_0 \left(\frac{\pi r^2 - \pi a^2}{\pi b^2 - \pi a^2} i \right)$$

$$B = \frac{\mu_0 i}{2\pi r} \left(\frac{r^2 - a^2}{b^2 - a^2} \right)$$

So,

$$B(r) = \begin{cases} 0, & \text{for } r < a \\ \dfrac{\mu_0 i}{2\pi r}\left(\dfrac{r^2 - a^2}{b^2 - a^2}\right), & \text{for } a \leq r < b \\ \dfrac{\mu_0(i)}{2\pi r}, & \text{for } r \geq b \end{cases}$$

EXAMPLE

A solenoid is a long tube with a wire wrapped around it (long means that the length is greater than the diameter of the tube). What is the magnetic field in the center of the solenoid? The solenoid has a length of L and the wire of current i is wrapped around the solenoid N times.

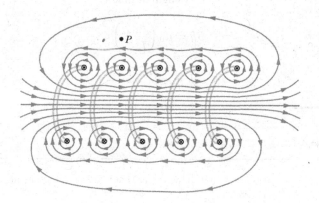

We want the field in the center, and the field runs along the length of the solenoid, so our path must run down the center of the solenoid. We choose a rectangle that goes a distance h down the axis of the solenoid, then out and back.

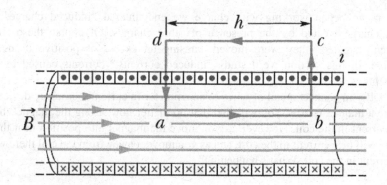

Down the axis, the magnetic field is constant and parallel to the path. On the up and down legs of the path, the magnetic field is perpendicular to the path, so the magnetic field component parallel to the path is zero. Outside the solenoid the magnetic field is very small (not quite zero but much smaller than inside the solenoid).

$$BL = \oint \boldsymbol{B} \cdot \text{d}\boldsymbol{s} = \mu_0 I_{\text{enc}}$$

$$\oint \boldsymbol{B} \cdot \text{d}\boldsymbol{s} = (Bh) + (0) + (0 \times h) + (0)$$

$$Bh = \oint \boldsymbol{B} \cdot \text{d}\boldsymbol{s} = \mu_0 I_{\text{enc}}$$

How much current is enclosed? Many "turns" or wraps of the wire go through the loop, and each carries a current of i.

$$I_{\text{enc}} = (\text{number of turns enclosed})\, i$$

$$I_{\text{enc}} = \left(\frac{(\text{length of tube enclosed})}{(\text{length of tube})} N \right) i$$

$$Bh = \mu_0 \left(\frac{h}{L} N \right) i$$

$$B = \frac{\mu_0 N i}{L}$$

LESSON

Like Gauss's law, Ampere's law is always true, but to use it to find the magnetic field we need a symmetric problem.

9.3 INDUCED CURRENTS

When we began working with charge we encountered "induced charge." This was charge created by the presence of other charges. Of course these charges weren't created—they were moved, causing an excess of positive or negative charge. In this section we'll study "induced currents," currents caused by other currents.

If you connect a coil of wires to an ammeter (a current measuring device) and place a magnet nearby, you will find that no matter how strong the magnet there is no current in the coil. However, as you move the magnet into position near the coil there will be a current in the coil, and as you move it away from the coil there will be a current in the coil. What's happening?

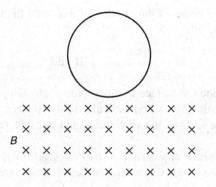

Imagine a loop of wire on the edge of a magnetic field. If the loop moves down, the positive charges along the bottom of the loop experience a force to the right. They push a current counterclockwise around the loop.

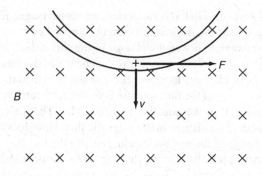

If the loop is entirely in the magnetic field (rather than partly in the magnetic field) then the positive charges at the top of the loop also feel a force to the right, and the result is no net current. This is similar to the magnet and coil, where the magnetic field itself doesn't cause a current but a movement in the magnetic field does.

The voltage induced in a loop by a magnetic field is

$$\mathcal{E} = -\frac{d}{dt}\Phi_B = -\frac{\Delta\Phi_B}{\Delta t}$$

where Φ_B is the magnetic field flux passing through the loop. This voltage causes a current if the loop is closed (complete). What is voltage? We'll explain it in the next couple of chapters, but for now it is what causes charges to move in a current.

What is flux? (If you remember this from Gauss's law then this is a repeat.) Imagine that you are out in the rain holding a bucket. Flux would be the rate at which raindrops enter the bucket. You could reduce this rate by using a smaller bucket, by going to a place with less rain, or by tilting the bucket sideways. The magnetic field flux is effectively a count of the number of magnetic field lines passing through a surface. Mathematically,

$$\Phi_B = \int \mathbf{B} \cdot d\mathbf{A}$$

which says to divide the surface into tiny pieces, at each piece find the component of the magnetic field that is perpendicular to the surface (normal to the surface), multiply times the area of the tiny piece, and add the results for all of the tiny pieces.

Despite the scary appearance of this integral, it's not so bad. Typically, the magnetic field is the same everywhere on the surface (or we'll pretend it is), so

constant field

$$\Phi_B = \int \mathbf{B} \cdot d\mathbf{A} = BA$$

definition of flux

The derivative (d) or delta (Δ) indicates that it is not magnetic field that creates an induced voltage or current. An induced current is caused by a *change* in the magnetic flux. In the first example it was the change in the flux caused by the movement of the magnet. In the second example it was the change in the flux as the loop moved into the magnetic field. If the loop is entirely in the field then movement of the loop doesn't cause a change in the flux so there is no induced current.

The minus sign in the equation is a reminder. The direction of the induced current is opposite to the change in the magnetic flux. How does this work? Consider our loop on the edge of the magnetic field. Initially the flux through the loop is zero since there is no magnetic field where the loop is. As we move the loop down, into the magnetic field, the magnetic flux into the page increases. The loop "wants" to keep the same flux it had (zero, in this case), so it creates magnetic field out of the page. That is, there will be an induced current in the direction that causes a magnetic field to pass through the loop out of the page.

Which direction would this current be? There are two ways to figure it out using the right-hand rules from the magnetic field section. One way is to point your thumb in the direction that you want the magnetic field through the loop to go (out of the page), then the fingers curl in the direction of the current (counterclockwise). The other is to point the fingers in the direction that you want the magnetic field through the loop to go. Then the thumb goes around the fingers in the direction of the current (counterclockwise).

EXAMPLE

A loop of wire is in the same plane as a long straight wire. The straight wire carries a constant current to the left. What is the direction of the current induced in the loop as it moves up? What is the direction of the current induced in the loop as it moves left?

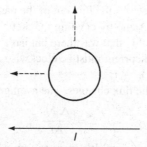

What is the direction of the magnetic field that the straight current causes at the loop? Take your right hand and point the thumb in the direction of the current (to the left). The fingers go away from you (into the paper) above the wire, and toward you (out of the paper) below the wire. The loop is above the wire, so the magnetic field there is into the page.

What happens to the magnetic flux as the loop moves up? When you go further away from the straight wire the magnetic field becomes weaker ($B = \mu_0 I / 2\pi R$). As the loop moves to where the magnetic field is weaker the flux decreases. The magnetic flux into the paper decreases, so the loop tries to make magnetic field into the paper, to replace the flux that was there but isn't anymore. To create magnetic field into the page the current must be clockwise (use one of the right-hand rules).

What happens to the magnetic flux as the loop moves to the left? It moves from one spot to another where the magnetic field is the same, so the flux doesn't change. If the flux doesn't change then there is no induced current.

EXAMPLE

A rectangular loop consists of 25 turns of wire, each 7 cm by 9 cm. It is oriented vertically in a plane in and out of the page. The 0.15 T magnetic field points to the left. As the loop rotates 90° about the horizontal axis in 0.025 seconds, what are the direction of the induced current and the average magnitude of the induced voltage?

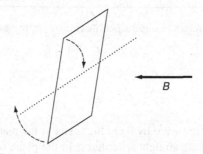

B

Before the loop rotates the magnetic flux is to the left. After the loop rotates 90° the flux is zero—the magnetic field goes by the loop but does not go through, instead skimming the surface. As the flux out of the page decreases, the loop tries to keep the magnetic flux the same by creating magnetic field to the left (up after the loop rotates) to replace the flux that was there but isn't anymore. To create magnetic field to the left, the induced current must be clockwise as seen from the right (as the loop is seen before rotating).

As the loop rotates, the flux changes. The average voltage induced is

$$\mathcal{E} = \frac{\Delta \Phi_B}{\Delta t}$$

What happened to the minus sign? It's there to remind us that the induced current is *opposite* to the change in the flux. We've already figured out the direction of the induced current, and now we're trying to find the magnitude. The magnitude isn't negative.

$$\mathcal{E} = \left| \frac{\Delta(BA)}{\Delta t} \right| = \left| \frac{(B_f A - B_i A)}{\Delta t} \right|$$

After the loop has rotated 90° the flux is zero—none of the magnetic field is going through the loop.

$$\mathcal{E} = \frac{B_i A}{\Delta t}$$

$$\mathcal{E} = \frac{(0.15 \text{ T}) [(7 \text{ cm})(9 \text{ cm})]}{(0.025 \text{ s})}$$

$$\mathcal{E} = 0.038 \text{ T m}^2/\text{s} = 0.038 \text{ V} = 38 \text{ mV}$$

This is the voltage induced in each loop. There are 25 loops connected together, so the voltage is

$$25 \times 0.038 \text{ V} = 0.95 \text{ V}$$

LESSON

The induced current creates a magnetic field that is opposite to the *change* in the magnetic flux.

EXAMPLE⋆

A rectangular loop (7 cm wide by 9 cm high) sits in the plane of the page. The top edge is 5 cm from a long straight wire that is in the plane of the loop. The current in the straight wire is 5 A and is increasing at a rate of 180 A/s. What is the voltage induced in the loop?

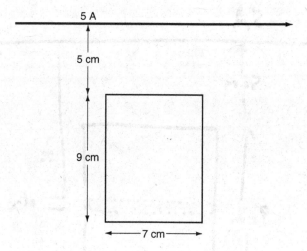

The induced voltage is

$$\mathcal{E} = \frac{d}{dt} \Phi_B$$

The magnetic flux is

$$\Phi_B = \int B \, dA$$

The magnetic field is not the same everywhere on the loop, so we can't use $\Phi_B = BA$. Instead we divide the surface formed by the loop into many little pieces, each so small that the magnetic field is the same everywhere on each piece. We use pieces 7 cm wide by dx high.

$$\Phi_B = \int B \, dA$$

$$\Phi_B = \int_{5 \text{ cm}}^{14 \text{ cm}} \frac{\mu_0 i}{2\pi x} \, [(7 \text{ cm}) \, dx]$$

$$\Phi_B = \frac{\mu_0 i (7 \text{ cm})}{2\pi} \int_{5 \text{ cm}}^{14 \text{ cm}} \frac{1}{x} \, dx$$

$$\Phi_B = \frac{\mu_0 i (7 \text{ cm})}{2\pi} \, [\ln x]_{5 \text{ cm}}^{14 \text{ cm}}$$

$$\Phi_B = \frac{\mu_0 i (7 \text{ cm})}{2\pi} \, [\ln (14 \text{ cm}) - \ln (5 \text{ cm})]$$

$$\Phi_B = \frac{\mu_0 i (7 \text{ cm})}{2\pi} \left[\ln \left(\frac{14 \text{ cm}}{5 \text{ cm}} \right) \right]$$

$$\Phi_B = \frac{\mu_0 i (7 \text{ cm})}{2\pi} \, \ln (2.8)$$

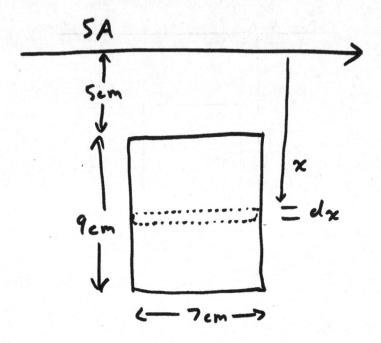

The induced voltage is

$$\mathcal{E} = \frac{\mathrm{d}}{\mathrm{d}t}\Phi_B$$

$$\mathcal{E} = \frac{\mathrm{d}}{\mathrm{d}t}\frac{\mu_0 i(7\text{ cm})}{2\pi}\ln(2.8)$$

$$\mathcal{E} = \left(\frac{\mathrm{d}i}{\mathrm{d}t}\right)\frac{\mu_0(7\text{ cm})}{2\pi}\ln(2.8)$$

$$\mathcal{E} = (180\text{ A/s})\frac{(4\pi\times 10^{-7}\text{ T m/A})(0.07\text{ m})}{2\pi}\ln(2.8)$$

$$\mathcal{E} = 2.6\times 10^{-6}\text{ T m}^2/\text{s} = 2.6\times 10^{-6}\text{ V} = 2.6\ \mu\text{V}$$

EXAMPLE

A small circular loop (8 cm diameter) of 10 turns sits in the plane of and inside of a large loop (28 cm diameter) of 100 turns. The current in the outer loop is 5 A and is increasing at a rate of 180 A/s. What is the voltage induced in the inner loop?

The induced *voltage* is

$$\mathcal{E} = \frac{\mathrm{d}}{\mathrm{d}t}\Phi_B$$

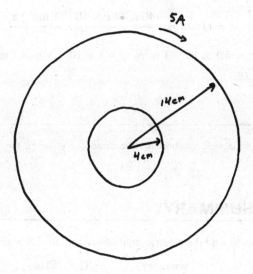

The magnetic flux is

$$\Phi_B = \int B \, dA$$

The magnetic field is not quite the same everywhere on the loop, but is reasonably close, so we'll use $\Phi_B = BA$.

$$\Phi_B = BA$$

The magnetic field is $B_{\text{loop}} = \mu_0 i / 2R$ from each of 100 loops of the larger coil.

$$\Phi_B = \left(100 \, \frac{\mu_0 i}{2R}\right) A$$

Only the field that passes through the smaller coil creates flux in the smaller coil.

$$\Phi_B = \left(100 \, \frac{\mu_0 i}{2(14 \text{ cm})}\right) \left[\pi(4 \text{ cm})^2\right]$$

$$\Phi_B = \frac{(4 \text{ m})\mu_0 i \, \pi}{7}$$

The induced voltage is (with 10 turns in the small coil)

$$\mathcal{E} = N_{\text{small}} \frac{d}{dt} \Phi_B$$

$$\mathcal{E} = (10) \frac{d}{dt} \frac{(4 \text{ m})\mu_0 i \, \pi}{7}$$

$$\mathcal{E} = \left(\frac{di}{dt}\right) \frac{(40 \text{ m})\mu_0 \, \pi}{7}$$

$$\mathcal{E} = (180 \text{ A/s}) \frac{(40 \text{ m})(4\pi \times 10^{-7} \text{ T m/A}) \pi}{7}$$

$$\mathcal{E} = 4.1 \times 10^{-3} \text{ T m}^2/\text{s} = 4.1 \times 10^{-3} \text{ V} = 4.1 \text{ mV}$$

LESSON

When the magnetic field is constant (the same everywhere), the magnetic flux is $\Phi_B = BA$.

CHAPTER SUMMARY

- Use Gauss's law to find the electric field of a symmetric charge distribution.

$$\overbrace{EA = \int E \cdot dA}^{\text{symmetry}} = \overbrace{\Phi_E = \frac{q_{net}}{\epsilon_0}}^{\text{Gauss's law}}$$

definition of flux

- Use Ampere's law to find the magnetic field of a symmetric current.

$$\overbrace{BL = \oint B \cdot ds = \mu_0 I_{\text{enc}}}^{\text{symmetry}}$$

Ampere's law

- A changing magnetic flux creates an induced current

$$\mathcal{E} = \frac{d}{dt}\Phi_B$$

where the magnetic flux is

$$\Phi_B = \int B \cdot dA = \overbrace{BA}^{\text{constant field}}$$

definition of flux

CHAPTER 10

POTENTIAL = VOLTAGE

We've learned to use forces when we deal with charged objects. Can we use conservation of energy and conservation of momentum?

Certainly energy and momentum are both conserved, but can we use them effectively to solve problems? Conservation of momentum doesn't help us because some of the charges are usually "glued down," and we don't know the force on them so we don't know the impulse. Conservation of energy is so effective in dealing with charges that we do it all the time, often without noticing.

How did we use conservation of energy earlier? The energy you start with plus the energy you put in (work) is the energy you end with.

$$KE + PE + W = KE' + PE'$$

The work done by some forces (gravity and springs) is easier to deal with by using potential energy, and then we don't include them in the work term. Work done by the electric force is the other work that is easier to do with potential energy.

How does potential energy from gravity work? If we start with a rock at the bottom of a hill and move it to the top of the hill, the work that gravity does during the process is the same no matter how we get the rock to the top. The work that gravity does is negative because the gravity force is down but the motion is up. We could get gravity to do positive work by letting the rock roll down the hill—that's why it's called potential energy. If we tied a string to the rock beforehand and wrapped the other end around a generator (next chapter), then as the rock rolled down the hill we could get electricity to power our stereo and make tunes, which is the purpose of physics.

The work that we do that is stored as potential energy is mgh. If we moved a rock that was twice as big then the potential energy would be twice as big. The potential energy is the product of the size of the rock (m) and something that depends on the position (gh). This latter part, the part that depends on position, is the **potential**. The units of gh are m^2/s^2 or J/kg—how much work it takes to move one kilogram of rock.

With electricity, the potential energy is equal to the product of the size (charge q) of the thing we move times the potential V, which depends on the position.

$$U = PE = qV$$

(U is often used for potential energy.) Potential is measured in joules per coulomb or volts, and is sometimes called voltage.

$$1 \text{ V} = \frac{1 \text{ J}}{1 \text{ C}}$$

Remember that potential and voltage are the same thing, but potential and potential energy are not the same. A location has potential, and a charge at the location has potential energy.

10.1 FINDING THE POTENTIAL

Potential energy is work that we've done that we might get back out. When we raise a rock, we exert a force equal to gravity mg (a little more at the beginning to get it going) while we raise it a height h. Since we push in the same direction as the motion we do positive work. Gravity pushes in the opposite direction as the motion and does negative work as the potential energy increases.

When we push a charge against the electrical force we do positive work, increasing the potential energy. When the charge moves on its own and we hold it back, we do negative work (or get work out) while the potential energy decreases (some of the potential energy is turned into work). To calculate the electrical potential energy we have to find the work we do pushing against the electric force, or negative the work done by the electric force.

$$\Delta PE = -\mathbf{F} \cdot \mathbf{d} = -q\mathbf{E} \cdot \mathbf{dd} = -qEd\cos\theta$$

Often the electric field is not constant as we move the charge. Therefore we divide the path into tiny pieces—so tiny that we can say that the electric field is constant for the whole tiny path. Then we add the work we've done (increase in potential energy) for all of the tiny pieces of the path.

$$\Delta PE = q \int -\mathbf{E} \cdot \mathrm{d}\mathbf{x}$$

where $\mathbf{E} \cdot \mathrm{d}\mathbf{x}$ is the component of the electric field parallel to the tiny piece $\mathrm{d}\mathbf{x}$ times the length of the tiny piece.

The potential energy is the product of two pieces, the amount of charge that we move and something that depends only on the path along which we move it. This second part is the potential.

$$\Delta V = \int -\mathbf{E} \cdot \mathrm{d}\mathbf{x}$$

EXAMPLE⋆

What is the potential difference between points C and D? What is the potential difference between C and E?

We have

$$\Delta V = \int -\mathbf{E} \cdot \mathrm{d}\mathbf{x}$$

or

$$V_D = V_C + \int_C^D -\mathbf{E} \cdot \mathrm{d}\mathbf{x}$$

$$V_D = V_C + \int_C^D -(-300 \text{ N/C } \hat{\mathbf{\imath}}) \cdot (\mathrm{d}x \, \hat{\mathbf{\imath}})$$

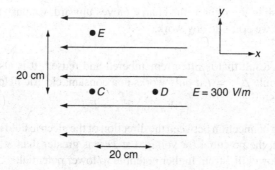

The electric field points in the negative x direction, while dx points in the positive x direction.

$$V_D = V_C + \int_C^D (300\ \text{N/C})(\hat{i} \cdot \hat{i})dx$$

The dot-product $\hat{i} \cdot \hat{i}$ means find the component of x that is parallel to x, so it's one.

$$V_D = V_C + (300\ \text{N/C}) \int_C^D dx$$

The integral is the sum of the lengths of the tiny pieces of the path, so it is the distance from C to D.

$$V_D = V_C + (300\ \text{N/C})(0.20\ \text{m})$$

$$V_D - V_C = 60\ \text{N m/C} = 60\ \text{J/C} = 60\ \text{V}$$

This says that D is 60 V higher in potential than C.

We can repeat the process for E.

$$V_E = V_C + \int_C^E -\mathbf{E} \cdot dx$$

$$V_E = V_C + \int_C^E -(-300\ \text{N/C}\ \hat{i}) \cdot (dx\ \hat{j})$$

The electric field points in the negative x direction, while dx points in the positive y direction.

$$V_E = V_C + \int_C^E (300\ \text{N/C})(\hat{i} \cdot \hat{j})dx$$

The dot-product $\hat{i} \cdot \hat{j}$ means find the component of x that is parallel to y, so it's zero.

$$V_E = V_C + (300\ \text{N/C})(0) \int_C^E dx$$

$$V_E = V_C$$

$$V_E - V_C = 0$$

Why is the potential difference between C and E zero? If we moved a positive charge from C to E, the electric force would be to the left and we would have to push right. If we push to the right as the charge moves upward, we push perpendicular to the motion and we don't do any work.

This is a result that is often remembered and reused. It is also misused quite often. The formula $\Delta V = Ed$ only works in a constant electric field.

$$\Delta V_{\text{constant field}} = Ed$$

There is a connection between the direction of the electric field and the potential. In the example, the potential (or voltage) at D was greater than at C. Electric field always runs "downhill," from higher potential to lower potential.

LESSON

Electric field points from higher potential to lower potential.

EXAMPLE*

What is the potential difference between C and D?

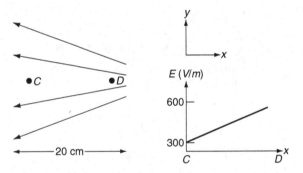

Again,

$$\Delta V = \int -E \cdot dx$$

$$V_D = V_C + \int_C^D -E \cdot dx$$

The electric field is not constant. The electric field magnitude is greatest when the density of field lines is greatest, or when the field lines are closest together. The electric field here is

$$E = \left[300 \text{ V/m} + (300 \text{ V/m})\left(\frac{x}{0.2 \text{ m}}\right)\right](-\hat{i})$$

*Example uses calculus.

Earlier the units for electric field were N/C. They are also equal to V/m.

$$E = \left[300 \text{ V/m} + (1500 \text{ V/m}^2)\, x\right](-\hat{\imath})$$

$$V_D = V_C + \int_C^D -\left[300 \text{ V/m} + (1500 \text{ V/m}^2)x\right](-\hat{\imath}) \cdot (dx\ \hat{\imath})$$

$$V_D = V_C - (-\hat{\imath} \cdot \hat{\imath}) \int_C^D \left[300 \text{ V/m} + (1500 \text{ V/m}^2)\, x\right] dx$$

$$V_D = V_C + \int_C^D \left[300 \text{ V/m} + (1500 \text{ V/m}^2)\, x\right] dx$$

$$V_D = V_C + \int_C^D (300 \text{ V/m})\, dx + \int_C^D (1500 \text{ V/m}^2)x\ dx$$

$$V_D = V_C + [(300 \text{ V/m})x]_0^{0.2 \text{ m}} + \left[(1500 \text{ V/m}^2)\frac{1}{2}x^2\right]_0^{0.2 \text{ m}}$$

$$V_D = V_C + (300 \text{ V/m})\left[(0.2 \text{ m}) - (0)\right] + \frac{1}{2}(1500 \text{ V/m}^2)[(0.2 \text{ m})^2 - (0)^2]$$

$$V_D = V_C + (60 \text{ V}) + (30 \text{ V})$$

$$V_D - V_C = 90 \text{ V}$$

This says that D is 90 V higher in potential than C.

LESSON

Only use $\Delta V = Ed$ in a constant electric field.

When we did potential energy with gravity, we could pick anywhere we liked as zero height as long as we used it consistently. This was because the initial and final potential energies appeared on opposite sides of the equation, and so only the difference in potential energy mattered. The same is true with electricity, so only the difference in potential matters. When we say the "potential at a point" we mean the difference in potential between that point and wherever we chose to be zero potential.

EXAMPLE★

What is the potential a distance R from a point charge q?

$$\Delta V = \int -E \cdot dx$$

Only the potential difference creates a physical effect. When we speak of the potential at a point we mean the potential difference compared to some other point. If that point is not given, then it is taken to be at infinity, or at an infinite distance from any other charge in the problem.

$$V_R = V_\infty + \int_\infty^R -\mathbf{E} \cdot \mathrm{d}\mathbf{x}$$

Personally, I hate doing integrals in the negative direction—I always seem to get a minus sign wrong.

$$V_\infty = V_R + \int_R^\infty -\mathbf{E} \cdot \mathrm{d}\mathbf{x}$$

V_∞ is the potential at ∞, which is zero.

$$0 = V_R + \int_R^\infty -\mathbf{E} \cdot \mathrm{d}\mathbf{x}$$

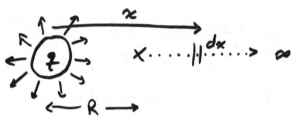

The electric field is kq/r^2 pointing away from the charge, always parallel to our path.

$$0 = V_R + \int_R^\infty -\left(\frac{kq}{r^2}\right) \mathrm{d}r$$

$$0 = V_R - kq \int_R^\infty \frac{\mathrm{d}r}{r^2}$$

$$0 = V_R - kq \left[-\frac{1}{r}\right]_R^\infty$$

$$0 = V_R - kq \left[\left(-\frac{1}{\infty}\right) - \left(-\frac{1}{R}\right)\right]$$

$$0 = V_R - kq \left(\frac{1}{R}\right)$$

$$V_R = \frac{kq}{R}$$

The result, the potential created by a point charge, is worth remembering. It is part of what makes potential worth using.

$$V_{\text{point charge}} = \frac{kq}{R}$$

One of the advantages of using conservation of energy was that energy doesn't have direction. We never had to find and add components. The same is true with electrical energy, so potential doesn't have direction. To find the electric field created by two charges, we had to add the electric fields as vectors. To find the potential created by two charges, we add the potentials created by each charge, but not as vectors.

EXAMPLE

What is the potential at point P created by the $+5\ \mu C$ and $-5\ \mu C$ charges?

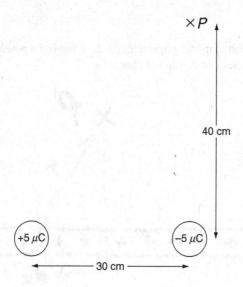

To use the equation:

$$\Delta V = \int -\boldsymbol{E} \cdot \mathrm{d}\boldsymbol{x}$$

we need to choose a path (any path) that goes from someplace where we know the potential (at ∞) to the point P. Then we need to find the electric field vector everywhere along this path. This is doable but involves nasty calculus. Part of the reason for using potential is to avoid calculus (really).

Instead of doing this integral, let's find the potential from each point charge and add the potentials.

$$V_P = V_{+5} + V_{-5}$$

$$V_P = \frac{k(+5\ \mu C)}{R_+} + \frac{k(-5\ \mu C)}{R_-}$$

$$V_P = \frac{k(+5\ \mu C)}{\sqrt{(30\ \mathrm{cm})^2 + (40\ \mathrm{cm})^2}} + \frac{k(-5\ \mu C)}{(40\ \mathrm{cm})}$$

$$V_P = \frac{k(+5\ \mu C)}{(50\ \text{cm})} + \frac{k(-5\ \mu C)}{(40\ \text{cm})}$$

$$V_P = \frac{(9 \times 10^9\ \text{N m}^2/\text{C}^2)(+5 \times 10^{-6}\ \text{C})}{(0.50\ \text{m})} + \frac{(9 \times 10^9\ \text{N m}^2/\text{C}^2)(-5 \times 10^{-6}\ \text{C})}{(0.40\ \text{m})}$$

$$V_P = (9 \times 10^4\ \text{N m/C}) + (-11.25 \times 10^4\ \text{N m/C})$$

$$V_P = -22500\ \text{V} = -22.5\ \text{kV}$$

EXAMPLE⋆

A charge Q is spread uniformly over a length L. What is the potential at P, a distance d away from the center of the line of charge?

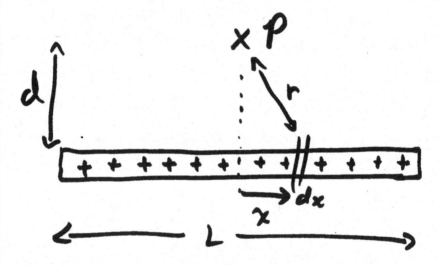

There are two ways to approach this problem. We could start at infinity, where we know the potential is zero, and move to P, adding the electric field. We could also divide the charge into tiny pieces, each so small that all of the charge in it is the same distance from P, and add the potentials created by all of the tiny pieces. To compare the two methods, we'll do both.

First we divide the line into little pieces of charge. The potential that each piece of the line creates at P is

$$V = \frac{k\ (\text{charge of the piece})}{(\text{distance to } P)}$$

We need to know the distance from each piece to the point P and the charge of each piece.

We'll use x for the distance from the midpoint of the line to a piece of the line. Thus, x will go from a minimum of $-L/2$ to a maximum of $+L/2$. The width of each piece is then dx, the distance from one piece of the line to the next. The distance from each piece to P is

$$r = \sqrt{d^2 + x^2}$$

We need to know the charge in each piece. If the size of the piece dx is $1/100$ of L then the charge is $1/100$ of Q. The size of each piece is dx/L of the length so the charge is dx/L of the charge, or $Q(dx/L)$. Alternatively, the density of charge along the line is $\lambda = Q/L$ and the charge in a piece of length dx is $(Q/L)dx$.

The potential at P is

$$V_P = \int_{-L/2}^{+L/2} \frac{k(Q\,dx/L)}{\sqrt{d^2 + x^2}}$$

$$V_P = \frac{kQ}{L} \int_{-L/2}^{+L/2} \frac{dx}{\sqrt{d^2 + x^2}}$$

$$V_P = \frac{kQ}{L} \left[\ln\left(x + \sqrt{d^2 + x^2}\right) \right]_{-L/2}^{+L/2}$$

$$V_P = \frac{kQ}{L} \left(\ln\left[(L/2) + \sqrt{d^2 + (L/2)^2} \right] - \ln\left[(-L/2) + \sqrt{d^2 + (-L/2)^2} \right] \right)$$

$$V_P = \frac{kQ}{L} \left[\ln\left(\frac{(L/2) + \sqrt{d^2 + (L/2)^2}}{(-L/2) + \sqrt{d^2 + (L/2)^2}} \right) \right]$$

$$V_P = \frac{kQ}{L} \ln\left(\frac{(L/2) + \sqrt{d^2 + (L/2)^2}}{(-L/2) + \sqrt{d^2 + (L/2)^2}} \cdot \frac{(L/2) + \sqrt{d^2 + (L/2)^2}}{(L/2) + \sqrt{d^2 + (L/2)^2}} \right)$$

$$V_P = \frac{kQ}{L} \ln\left(\frac{\left((L/2) + \sqrt{d^2 + (L/2)^2}\right)^2}{-(L/2)^2 + (\sqrt{d^2 + (L/2)^2})^2} \right)$$

$$V_P = \frac{kQ}{L} \ln\left(\frac{\left((L/2) + \sqrt{d^2 + (L/2)^2}\right)^2}{d^2} \right)$$

$$V_P = \frac{2kQ}{L} \ln\left(\frac{(L/2) + \sqrt{d^2 + (L/2)^2}}{d} \right)$$

The other method is to integrate the electric field in from infinity.

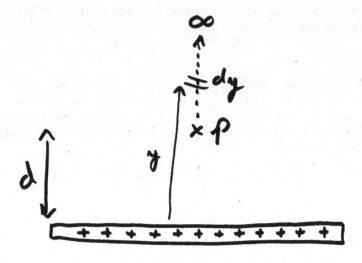

$$V_P = V_\infty + \int_\infty^P -\mathbf{E} \cdot \mathrm{d}\mathbf{x}$$

Instead I'll start at P and go out to infinity.

$$V_\infty = V_P + \int_P^\infty -\mathbf{E} \cdot \mathrm{d}\mathbf{x}$$

I'll follow a path straight up along the y-axis. The electric field points this way, so the electric field is parallel to each tiny step. When we get to infinity the potential is zero.

$$0 = V_P + \int_P^\infty -E_y \, \mathrm{d}y$$

We need to find the electric field at each spot along the y-axis. Fortunately we did this in an example a couple chapters ago.

$$E_y(y) = \frac{kQ}{y\sqrt{y^2 + (L/2)^2}}$$

$$0 = V_P + \int_d^\infty -\frac{kQ}{y\sqrt{y^2 + (L/2)^2}} \, \mathrm{d}y$$

$$V_P = kQ \int_d^\infty \frac{\mathrm{d}y}{y\sqrt{y^2 + (L/2)^2}}$$

$$V_P = kQ \left[-\frac{1}{(L/2)} \ln\left(\frac{(L/2) + \sqrt{y^2 + (L/2)^2}}{y} \right) \right]_d^\infty$$

$$V_P = \frac{2kQ}{L} \left[-\ln\left(\frac{(L/2) + \sqrt{\infty^2 + (L/2)^2}}{\infty} \right) + \ln\left(\frac{(L/2) + \sqrt{d^2 + (L/2)^2}}{d} \right) \right]$$

$$V_P = \frac{2kQ}{L}\left[-\ln\left(\frac{(L/2)+\infty}{\infty}\right) + \ln\left(\frac{(L/2)+\sqrt{d^2+(L/2)^2}}{d}\right)\right]$$

$$V_P = \frac{2kQ}{L}\ln\left(\frac{(L/2)+\sqrt{d^2+(L/2)^2}}{d}\right)$$

It is true that the second method appears shorter here. That is partly because we did more algebra after the integration on the first method but mostly because the first part of the second method had already been done. If you have a choice, first try to use superposition—adding the potentials created by each piece of the charge.

LESSON

For multiple charges, either find the potential for each charge and add them (superposition) or follow a path from infinity, integrating the electric field.

10.2 USING THE POTENTIAL

When we used conservation of energy, it looked like

$$E_{\text{before}} + W = E_{\text{after}}$$

Then we introduced potential energy. Some forces exert the same work on something as it moves from A to B no matter what path the object takes. The work these forces do can be represented with potential energy

$$KE + PE + W = KE' + PE'$$

If we treat a force with potential energy then we don't include the work done by that force—potential energy does it. Gravity and springs could be treated with potential energy. The third such force is electric force (but not magnetic force, which never does any work because the force is perpendicular to the motion). The potential energy of electricity is

$$U = PE = qV$$

EXAMPLE

An electron starts at point A ($V_A = 20$ V) at a speed of 2.5×10^6 m/s. How fast is it going when it gets to spot B ($V_B = 30$ V)?

We might think of using forces, finding the electric field, then the electric force, then the work done as the electron moves each little bit of the path. The purpose of potential is to relieve us of this task. Besides, we don't know

anything about the charges that create the electric field. Instead we use conservation of energy.

$$KE + PE + W = KE' + PE'$$

$$\frac{1}{2}mv^2 + qV_A + (0) = \frac{1}{2}m(v')^2 + qV_B$$

$$\frac{1}{2}m(v')^2 = \frac{1}{2}mv^2 + qV_A - qV_B$$

$$\frac{1}{2}m(v')^2 = \frac{1}{2}mv^2 + q(V_A - V_B)$$

$$\frac{1}{2}m(v')^2 = \frac{1}{2}mv^2 + (-e)(V_A - V_B)$$

$$(v')^2 = v^2 - \frac{2e}{m}(V_A - V_B)$$

$$v' = \sqrt{(2.5 \times 10^6 \text{ m/s})^2 - \frac{2(1.6 \times 10^{-19} \text{ C})}{(9.1 \times 10^{-31} \text{ kg})}[(20 \text{ V}) - (30 \text{ V})]}$$

$$v' = 3.1 \times 10^6 \text{ m/s}$$

How does the electron speed up when it moves to a *higher* potential? Because the electron has negative charge, it has a more negative or lower potential energy at 30 V than it does at 20 V.

EXAMPLE

An electric field points from left to right. An electron is placed in the electric field and set free. Which way does the electron go?

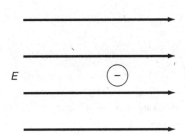

There are lots of ways to think about this. First, the electric field comes from positive charges and goes to negative charges. There must be positive charges to the left and negative charges to the right (out of view—pay no attention to those charges behind the curtain). The electron is attracted to the positive charges and repelled by the negative charges, to it moves to the left. Also, the force on a negative charge is in the opposite direction as the electric field, and the field points to the right, so the force on the electron is to the left.

Another way to approach the problem is to use conservation of energy.

$$KE + PE + W = KE' + PE'$$

The electron will go whichever way it gains speed. It starts with no kinetic energy. The only force is electricity, which we'll treat with potential energy instead of work.

$$PE = KE' + PE'$$

$$KE' = PE - PE'$$

$$KE' = qV - qV'$$

$$KE' = (-e)V - (-e)V'$$

$$KE' = e\left(V' - V\right)$$

The electron moves to a higher potential, so that $V' > V$ and $(V' - V)$ is positive. Since the electric field points from higher potential to lower potential, the electron moves to the left.

For a proton (with a positive charge),

$$KE' = qV - qV'$$

$$KE' = (+e)V - (+e)V'$$

$$KE' = e\left(V - V'\right)$$

The proton moves to a lower potential, so that $V' < V$ and $(V - V')$ is positive. The proton moves to the right.

LESSON

Everything moves to where it has less potential energy (and more kinetic energy). This may not be the same as less potential.

EXAMPLE

How much work can an electron do in moving between the terminals of a 1.5 V battery?

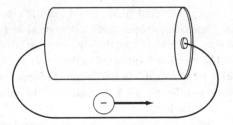

We use conservation of energy.

$$KE + PE + W = KE' + PE'$$

The charges have no kinetic energy to start with, so $KE = 0$. The more kinetic energy they have at the end the less work they could have done, so $KE' = 0$. The potential energies are $PE = qV$.

$$qV + W = qV'$$

$$W = qV' - qV$$

$$W = q(V' - V)$$

Electrons move from the negative or lower voltage side of a battery to the higher end, so the final potential is higher than the initial potential. The difference in the potentials or voltages is 1.5 V—a battery is a device to create a constant potential difference.

$$W = (-e)(1.5 \text{ V})$$

$$W = (-1.6 \times 10^{-19} \text{ C})(1.5 \text{ V})$$

$$W = -2.4 \times 10^{-19} \text{ J}$$

This is the work done on the electron. The work done by the electron is $+2.4 \times 10^{-19}$ J.

A coulomb of charge is $1/(1.6 \times 10^{-19} \text{ C}) = 6 \times 10^{18}$ electrons, so the work that a coulomb of charge does is

$$\left(6 \times 10^{18}\right) \times \left(2.4 \times 10^{-19} \text{ J}\right) = 1.5 \text{ J}$$

A coulomb is a big charge, and sometimes we have only small charges like a single electron. When lots of charge moves there is a lot of energy involved,

$$1 \text{ C} \times 1 \text{ V} = 1 \text{ J}$$

but when only a little charge moves there is a small amount of energy involved,

$$1 e \times 1 \text{ V} = 1 \text{ eV}$$

where eV is short for "electron-volt."

EXAMPLE

What is the kinetic energy that an electron acquires as it moves from ground to 300 V? What is the kinetic energy that an electron acquires as it moves from ground to −300 V?

$$? \leftarrow \ominus \rightarrow ?$$

$$\times \qquad \times \qquad \times$$

$$-300\,V \qquad 0V \qquad 300\,V$$

Earlier "ground" was a large conductor, so large that it was always neutral. When using potential, "ground" is the name given to the place where potential is zero ($V_{ground} = 0$). You have to tell which way the word is being used from the context.

We use conservation of energy.

$$KE + PE + W = KE' + PE'$$

$$0 + (-e)(0) + 0 = KE' + (-e)(300\text{ V})$$

$$KE' = (e)(300\text{ V}) = 300\text{ eV}$$

When it moves from ground to a location where the potential is −300 V,

$$KE + PE + W = KE' + PE'$$

$$0 + (-e)(0) + 0 = KE' + (-e)(-300\text{ V})$$

$$KE' = -(e)(300\text{ V}) = -300\text{ eV}$$

How can the kinetic energy be negative? The electron won't go there on its own—we would have to push it, doing work, to get the electron to go from ground to −300 V.

EXAMPLE

What is the work needed to move a proton from P to Q?

We use conservation of energy.

$$KE + PE + W = KE' + PE'$$

$$0 + (+e)V_P + W = 0 + (+e)V_Q$$

$$W = (+e)V_Q - (+e)V_P$$

We need to find the potentials at P and Q. We found the potential at P in an earlier example.

$$V_P = -22500\text{ V} = -22.5\text{ kV}$$

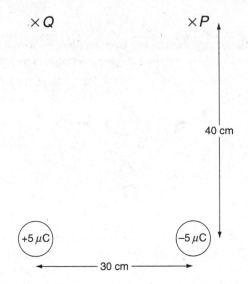

We find the potential at Q the same way.

$$V_Q = V_{+5} + V_{-5}$$

$$V_Q = \frac{k(+5\ \mu C)}{(40\ \text{cm})} + \frac{k(-5\ \mu C)}{(50\ \text{cm})}$$

$$V_Q = \frac{(9 \times 10^9\ \text{N m}^2/\text{C}^2)(+5 \times 10^{-6}\ \text{C})}{(0.40\ \text{m})} + \frac{(9 \times 10^9\ \text{N m}^2/\text{C}^2)(-5 \times 10^{-6}\ \text{C})}{(0.50\ \text{m})}$$

$$V_Q = 22500\ \text{V} = 22.5\ \text{kV}$$

$$W = (+e)\left(V_Q - V_P\right)$$

$$W = (+e)[(+22500\ \text{V}) - (-22500\ \text{V})]$$

$$W = (+1.6 \times 10^{-19}\ \text{C})(+45000\ \text{V})$$

$$W = +7.2 \times 10^{-15}\ \text{J}\ \text{ or }\ +45000\ \text{eV}$$

If the work were negative, then the charge would have gone from P to Q on its own.

LESSON

To use potential, use conservation of energy with $PE = qV$.

CHAPTER SUMMARY

- Use conservation of energy with $PE = qV$.

- Everything tries to move to where it has less potential energy. For negative charges this is at higher potential.

- The potential difference between two spots is

$$\Delta V = \int -\boldsymbol{E} \cdot \mathrm{d}\boldsymbol{x}$$

For a constant electric field this is

$$\Delta V_{\text{constant field}} = Ed$$

For a point charge this is

$$V_{\text{point charge}} = \frac{kq}{R}$$

- For multiple charges, try to use superposition, finding the potential created by each charge and adding. Potential does not have direction, so don't add potentials as vectors.

CHAPTER 11

CIRCUITS

In the last chapter we saw that as a charge goes from a higher potential energy to a lower one, electric forces do work on it. It is possible for this work to be turned into other forms of energy. While a single charge does not do much work, a current of many charges can do a substantial amount. Eventually we'd like to turn this work into tunes or some other useful form of energy.

First we'll focus on simple circuits involving resistors or capacitors and a single battery. The technique for dealing with these circuits is straightforward if overly repetitive. As usual when speaking physics, it's a matter of being clear in your language, in this case avoiding pronouns.

More advanced circuits require more advanced tools. Using the ideas we develop with the simpler circuits, we'll tackle some more interesting examples.

11.1 RESISTORS AND CAPACITORS

If we connect the two ends of a battery with a conducting wire, there will be a current from the higher potential end to the lower potential end. Current is really negative electrons moving from a lower potential to a higher potential. Instead we pretend that it's positive charges moving from higher potential to a lower potential.

What determines how much current? The property of materials to resist current is resistivity (ρ), and devices made of resistive material have resistance (R) and are called resistors. (We could use more original names, but they'd be harder to remember.) In a diagram, resistors are drawn as zigzag lines.

An electric field in conducting material causes a current. An electric field also causes a change in potential. It follows that the current through a resistor is connected to the voltage difference between the two sides. This connection is **Ohm's law**.

$$V = IR$$

The resistance is measured in ohms. To avoid confusing a capital "O" with a zero, the symbol for ohms is the Greek letter uppercase omega (1 Ω = 1 V/A).

Consider also the energy used in the resistor. Each charge uses an energy that depends on the change in the potential. If the voltage difference between the two sides of the resistor is 6 volts, or 6 joules per coulomb, then each coulomb of charge uses 6 joules of energy getting through the resistor. The more current, the faster coulombs of charge go through, and the quicker energy is turned into heat in the resistor. So the power, or rate of energy used in the resistor, is

$$P = IV$$

We have four variables (R, V, I, and P) and two equations. If we know any two of the four variables, we can find the other two using our two equations.

EXAMPLE

A lightbulb uses 100 watts of power when attached to 120 volts. What is the resistance of the lightbulb?

We find the resistance from Ohm's law using the voltage and the current. We don't know the current, but we can find it from $P = IV$.

$$V = IR \quad \text{and} \quad P = IV$$

$$R = \frac{V}{I} = \frac{V}{P/V} = \frac{V^2}{P}$$

$$R = \frac{(120 \text{ V})^2}{(100 \text{ W})} = 144 \ \Omega$$

EXAMPLE

A power plant delivers 100 MW of electrical power to the big city at 50 kV. The power travels through wires with a total resistance of 2 Ω. How much power is lost in the wires?

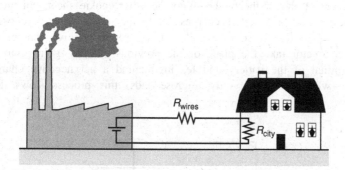

We have two resistors in this problem, the wires and the city. Normally wires don't have enough resistance for us to worry about, but the wires from the power plant are so long that their resistance is appreciable. Here 50 kV is the voltage difference between the two sides of the resistance of the city R_{city}. Likewise

100 MW is the power in R_{city}. The only value we know for the wires is their resistance R_{wire}.

	Wire	City
V		50 kV
I	$\leftrightarrow$	
R	2 Ω	
P	?	100 MW

The current that gets to the city goes through the wire, so the currents must be equal. If we use the information to find the I_{city}, that is equal to I_{wire}, and then we know two of the four for the wire.

$$P_{\text{city}} = I_{\text{city}} V_{\text{city}}$$

$$100 \text{ MW} = I_{\text{city}}(50 \text{ kV})$$

$$I_{\text{city}} = \frac{100 \text{ MW}}{50 \text{ kV}} = \frac{100 \times 10^6 \text{ W}}{50 \times 10^3 \text{ V}} = 2000 \text{ A}$$

$$P_{\text{wire}} = I_{\text{wire}} V_{\text{wire}} = I_{\text{wire}} (I_{\text{wire}} R_{\text{wire}}) = I_{\text{wire}}^2 R_{\text{wire}}$$

$$P_{\text{wire}} = (2000 \text{ A})^2 (2 \text{ } \Omega) = 8 \times 10^6 \text{ W} = 8 \text{ MW}$$

Eight percent of the power is lost in the wires. To avoid this, the electric company delivers electricity at much higher voltages, like 350 kV. At seven times the voltage, the current is one-seventh as much, and the power in the wires is 1/49 as much or 98% less—only 0.2% is lost.

Imagine taking two large flat pieces of metal, such as cookie sheets, and placing them parallel to each other but not touching. Then connect the two ends of a battery to the two cookie sheets. Positive charge flows out of the positive end of the battery to the first cookie sheet (really it's negative charges flowing out of the negative end of the battery onto the other cookie sheet, but the effect is the same). When the positive charges get to the first conductor (cookie sheet), they can go no further so they collect there. Other positive charges from the second conductor take the place of the previous charges, going into the negative terminal of the battery and leaving behind a net negative charge. After a while (which might be only microseconds) this process slows down and stops.

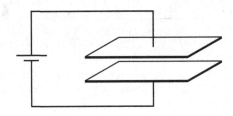

What if we were now to disconnect the battery from the cookie sheets? Could the charges on the conductors foresee what we're about to do and rush back into the battery? No, the charges are trapped on the conductors. We could now connect the conductors to the two ends of a lightbulb (or a resistor), and the excess positive charge would go through the bulb, creating a momentary current, and meet up with the excess negative charge on the other conductor. We have a device for storing small amounts of charge. The amount of charge Q that can be stored on a capacitor with capacitance C is

$$Q = CV$$

Equal amounts of positive and negative charge are stored on the two plates of the capacitor, and Q is the absolute value of one of these two charges. The units of capacitance are farads (F) and the symbol for a capacitor is two parallel lines of equal length. Compare this to the symbol for a battery, which is two parallel lines of unequal length (the longer one is always the higher potential).

It takes work to store charge on a capacitor. Since this energy can all be extracted later, we can express it as a potential energy

$$U_{\text{CAP}} = \frac{1}{2}QV$$

This is similar to the equation we had for the potential energy of a charge ($U = qV$) except for the one-half. One way to understand this difference is to imagine stacking books. If you start with a bunch of (equally sized) books on the floor, it doesn't take any work to put the first book on the floor. The second book takes a little work, since it must be moved up by the thickness of the first book. The third book takes even more work, and so on, until the last book takes the most work. We could add the work needed for each book, or multiply the number of books by the *average* work for each book. The average is the work for the middle book, which is lifted to half the height of the pile. Likewise it takes no work to put the first charge on the capacitor because with no charge it has no voltage. The second charge takes some work, the third more, and the last charge takes the most work. The total work is the charge times the average voltage, which is half of the final voltage. Use $U = qV$ for individual charges, but use $U = \frac{1}{2}QV$ when each successive charge takes more work than the next.

Just as we did with resistors, we have four variables (C, V, Q, and U) and two equations. If we know any two of the four variables, we can find the other two using our two equations. For both resistors and capacitors, we have the characteristic of the device, the voltage difference between the two ends, the charge (or charge per time = current), and the energy (or energy per time = power).

EXAMPLE

What is the energy that can be stored in a 0.10 μF capacitor that has a maximum voltage of 16 V?

The energy is

$$U = \frac{1}{2}QV = \frac{1}{2}(CV)V = \frac{1}{2}CV^2$$

$$U = \frac{1}{2}(0.10 \ \mu\text{F})(16 \ \text{V})^2 = 12.8 \ \mu\text{J}$$

11.2 SERIES AND PARALLEL

What happens if you have two (or more) resistors in the same circuit, such as in the circuit below? Charge can't collect at any point along the circuit—it would very quickly create a huge electric field. Therefore the current through the two resistors must be the same. Also, the sum of the voltages on the two resistors must be 12 V; that is, each coulomb of charge uses some of its 12 J of energy to get through the first resistor and the rest to get through the second resistor.

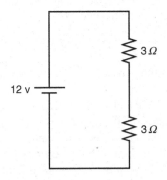

Since they have the same resistance and the same current, they must have the same voltage. These two voltages add up to 12 V. Two numbers that are the same and add up to twelve must be 6 V and 6 V. Each coulomb of charge uses 6 J of energy to get through the first resistor and 6 J of energy to get through the second resistor. The voltage difference between the two ends of each resistor is 6 V. The current through each resistor is

$$I = \frac{V}{R} = \frac{6 \ \text{V}}{3 \ \Omega} = 2 \ \text{A}$$

What if the two resistors are not equal? Let the bottom resistor be 6 Ω. The currents through the two resistors are still the same—the same as each other, not the same as before. Note this very important point: *if you change the circuit you change the currents.* The voltage across the top resistor is $V_3 = I(3 \ \Omega)$ and the voltage across

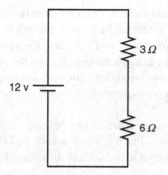

the bottom resistor is $V_6 = I(6\ \Omega)$. The sum of the two voltages is 12 V, so

$$12\ \text{V} = V_3 + V_6 = I(3\ \Omega) + I(6\ \Omega) = I(9\ \Omega)$$

$$I = \frac{12\ \text{V}}{9\ \Omega} = 1.3\ \text{A}$$

The voltage across the bottom resistor (potential difference between the two ends) is

$$V_6 = I_6 R_6 = (1.3\ \text{A})(6\ \Omega) = 8\ \text{V}$$

We could use a shortcut as in the previous problem. The currents are the same, and the 6 Ω resistor has twice the resistance as the 3 Ω resistor, so it must have twice the voltage. Can we think of two numbers where one is twice the other and the sum is 12? The voltage on the top resistor is 4 V and that on the bottom resistor is 8 V. (Usually the numbers don't come out so nice as they have here.)

When two devices are connected so that they must have the same current, we say they are in series. Two resistors in series act the same as a single resistor that has resistance

$$R_S = R_1 + R_2$$

Likewise when two devices are connected so that they must have the same voltage, we say they are in parallel. Two resistors in parallel act the same as a single resistor that has resistance

$$\frac{1}{R_P} = \frac{1}{R_1} + \frac{1}{R_2}$$

These rules are reversed for capacitors:

$$\frac{1}{C_S} = \frac{1}{C_1} + \frac{1}{C_2} \quad \text{and} \quad C_P = C_1 + C_2$$

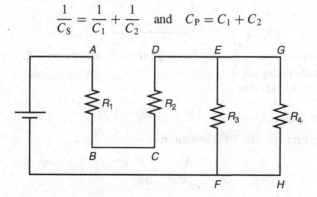

How do we determine whether two resistors (or capacitors) are in series? Consider R_1 and R_2 above. Current passes from A through R_1 to B. From there the charges can only go to C and through R_2 to D. R_1 and R_2 are in series because the charges that pass through R_1 must then go through R_2, and the currents through the two must be equal. Charges that pass through R_2 to point D can go either through R_3 to F or through R_4 to H. The currents through R_2 and R_3 do not have to be equal, so R_2 and R_3 are not in series.

How do we determine whether two resistors (or capacitors) are in parallel? Consider R_3 and R_4 above. Place a finger at each end of R_3. The voltage difference between your fingers is equal to the voltage difference between the ends of R_3, or the voltage across R_3. Now try moving your fingers, but only along wires and not across any circuit element, such as a resistor, capacitor, or battery. If by doing this you can get your fingers to lie at the two ends of R_4, then R_3 and R_4 have the same voltage across them and they are in parallel (they are, and you should be able to do this). Are R_3 and R_2 in parallel? We can move a finger from E to D but we can't get from F to C. We could only do this by jumping from one wire to another nearby wire (not allowed) or by going through the battery and R_1 (also not allowed). Therefore, R_3 and R_2 are not in parallel. Even though both are drawn vertically, so that they are geometrically parallel, they are not electrically parallel.

We have found that R_2 and R_3 are neither in series nor in parallel. Any two resistors do not have to be one or the other.

EXAMPLE

Find the current through and voltage across the 4 Ω resistor.

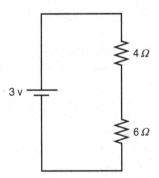

All of the charge that goes through the 4 Ω resistor must then go through the 6 Ω resistor, so they are in series.

Their combined resistance is

$$4 \,\Omega + 6 \,\Omega = 10 \,\Omega$$

The current through the 10 Ω resistor pair is

$$I = \frac{V}{R} = \frac{3 \text{ V}}{10 \,\Omega} = 0.30 \text{ A}$$

This current goes through both the 4 Ω and the 6 Ω resistors, so the current through the 4 Ω resistor is 0.30 A and the voltage difference between the two sides of the 4 Ω resistor is

$$V = IR = (0.30 \text{ A})(4 \text{ } \Omega) = 1.2 \text{ V}$$

EXAMPLE

Find the current through and voltage across the 4 Ω resistor and the current through the battery.

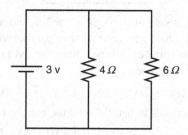

Put a finger on each side of the 4 Ω resistor, and move them left until they hit the battery terminals – the 4 Ω resistor is in parallel with the battery. Therefore they must have the same voltage, and the voltage across the 4 Ω resistor is 3 V. The current through the 4 Ω resistor is

$$I_4 = \frac{V_4}{R_4} = \frac{3 \text{ V}}{4 \text{ } \Omega} = 0.75 \text{ A}$$

Likewise the 6 Ω resistor is in parallel with the battery, has 3 V across it, and has

$$I_6 = \frac{V_6}{R_6} = \frac{3 \text{ V}}{6 \text{ } \Omega} = 0.50 \text{ A}$$

of current through it. The battery current is the sum of the resistor currents

$$I_\varepsilon = I_4 + I_6 = 0.75 \text{ A} + 0.50 \text{ A} = 1.25 \text{ A}$$

Alternatively, the resistors are in parallel (start with the fingers on the 4 Ω resistor and move them along only wires until they reach the ends of the 6 Ω resistor).

$$\frac{1}{R_{\text{pair}}} = \frac{1}{4 \text{ } \Omega} + \frac{1}{6 \text{ } \Omega} = \frac{6 \text{ } \Omega}{24 \text{ } (\Omega)^2} + \frac{4 \text{ } \Omega}{24 \text{ } (\Omega)^2}$$

$$R_{\text{pair}} = \frac{24 \text{ } (\Omega)^2}{10 \text{ } \Omega} = 2.4 \text{ } \Omega$$

$$I_\varepsilon = \frac{V}{R_{\text{pair}}} = \frac{3 \text{ V}}{2.4 \text{ } \Omega} = 1.25 \text{ A}$$

EXAMPLE

Find the charge on, and voltage across, the 4 μF capacitor and the total charge that flows from the battery.

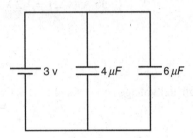

Put a finger on each side of the 4 μF capacitor, and move them left until they hit the battery terminals. The 4 μF capacitor is in parallel with the battery. Therefore they must have the same voltage, and the voltage across the 4 μF capacitor is 3 V. The charge on the 4 μF capacitor is

$$Q_4 = C_4 V_4 = (4 \ \mu F)(3 \ V) = 12 \ \mu C$$

Likewise the 6 μF capacitor is in parallel with the battery, has 3 V across it, and has

$$Q_6 = C_6 V_6 = (6 \ \mu F)(3 \ V) = 18 \ \mu C$$

of charge on it. The total charge that flows from the battery is the sum of the capacitor charges, with some of the charge going onto each capacitor.

$$Q_\varepsilon = Q_4 + Q_6 = 12 \ \mu C + 18 \ \mu C = 30 \ \mu C$$

Alternatively, the capacitors are in parallel (start with the fingers on the 4 μF capacitor and move them along only wires until they reach the ends of the 6 μF capacitor).

$$C_{\text{pair}} = 4 \ \mu F + 6 \ \mu F = 10 \ \mu F$$

$$Q_\varepsilon = C_{\text{pair}} V_{\text{pair}} = (10 \ \mu F)(3 \ V) = 30 \ \mu C$$

EXAMPLE

Find the charge on and voltage across the 4 μF capacitor.

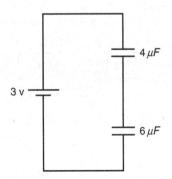

All of the charge that goes through the 4 μF capacitor must then go through the 6 μF capacitor, so they are in series. Their combined capacitance is

$$\frac{1}{C_{\text{pair}}} = \frac{1}{4\ \mu\text{F}} + \frac{1}{6\ \mu\text{F}} = \frac{6\ \mu\text{F}}{24\ (\mu\text{F})^2} + \frac{4\ \mu\text{F}}{24\ (\mu\text{F})^2}$$

$$C_{\text{pair}} = \frac{24\ (\mu\text{F})^2}{10\ \mu\text{F}} = 2.4\ \mu\text{F}$$

The charge on the 2.4 μF capacitor pair is

$$Q_{\text{pair}} = C_{\text{pair}} V_{\text{pair}} = (2.4\ \mu\text{F})(3\ \text{V}) = 7.2\ \mu\text{C}$$

This charge goes through both the 4 μF and the 6 μF capacitors, so the charge on the 4 μF capacitor is 7.2 μC and the charge on the 6 μF capacitor is 7.2 μC. The voltage difference between the two sides of the 4 μF capacitor is

$$Q = CV$$

$$7.2\ \mu\text{C} = (4\ \mu\text{F})V$$

$$V = 1.8\ \text{V}$$

LESSON

Series means same current; parallel means same voltage.

11.3 ONE-BATTERY CIRCUITS

When doing more complicated circuits, we often do the series and parallel thing many times. Ohm's law can be applied to any resistor, but it can also be applied to a pair in series, or to any group of resistors. Likewise $Q = CV$ can be applied to any capacitor, but it can also be applied to a pair in series, or to any group of capacitors. It's important to keep track of what value applies to what circuit element or group of elements.

EXAMPLE

What is the current through the 10 Ω resistor?

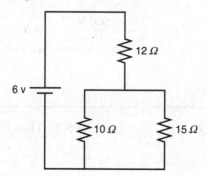

Is the voltage across the 10 Ω resistor equal to the battery voltage (6 V)? Put a finger on each side of the resistor and try to move them along wires, not crossing any circuit elements, until they lie on the two sides of the battery. You can get from the lower end of the resistor to the lower end of the battery, but to do so on the upper end you need to cross the 12 Ω resistor, which you may not do. The 10 Ω resistor is not in parallel with the battery, and they don't necessarily have the same voltage across them.

Is the 10 Ω resistor in series with the 12 Ω resistor? Current that passes through the 12 Ω resistor could go through the 10 Ω resistor but it could go through the 15 Ω resistor, so the 10 Ω and 12 Ω resistors are not in series.

Is the 10 Ω resistor in parallel with the 15 Ω resistor? Repeat the finger test, starting at the two ends of the 10 Ω resistor. You can move the fingers to the two ends of the 15 Ω resistor by moving only along wires, without crossing any circuit elements. They are in parallel, and we can replace them with a single resistor.

$$\frac{1}{R_{\text{pair}}} = \frac{1}{10\ \Omega} + \frac{1}{15\ \Omega} = \frac{15\ \Omega}{150\ (\Omega)^2} + \frac{10\ \Omega}{150\ (\Omega)^2}$$

$$R_{\text{pair}} = \frac{150\ (\Omega)^2}{25\ \Omega} = 6.0\ \Omega$$

This pair is in series with the 12 Ω resistor—current that goes through the 12 Ω resistor must go through the pair, passing through either the 10 Ω resistor or the 15 Ω resistor. The 12 Ω resistor and the 10/15 pair can be replaced with a single resistor.

$$R_{\text{trio}} = R_{12} + R_{\text{pair}} = 12\ \Omega + 6\ \Omega = 18\ \Omega$$

The current through the trio is the same as the current through the battery:

$$I_{\mathcal{E}} = I_{\text{trio}} = \frac{V_{\text{trio}}}{R_{\text{trio}}} = \frac{6\ \text{V}}{18\ \Omega} = 0.33\ \text{A}$$

All of the current passing through the trio passes through the 12 Ω resistor, so $I_{12} = I_{\text{trio}}$, but only some of this current goes through the 10 Ω resistor. We can find the voltage on the 10/15 pair.

$$V_{\text{pair}} = I_{\text{pair}} R_{\text{pair}} = (0.33\ \text{A})(6\ \Omega) = 2.0\ \text{V}$$

Because the 10 Ω resistor and the 15 Ω resistor are in parallel,

$$V_{10} = V_{15} = V_{\text{pair}} = 2.0\ \text{V}$$

Now that we know the voltage across the 10 Ω resistor we can find the current through it.

$$I_{10} = \frac{V_{10}}{R_{10}} = \frac{2.0\ \text{V}}{10\ \Omega} = 0.20\ \text{A}$$

LESSON

Use the finger test to check for parallel. Two things don't have to be one or the other.

It may seem strange, in the previous example, that $V_{12} + V_{10} + V_{15} = 4\,\text{V} + 2\,\text{V} + 2\,\text{V} = 8\,\text{V} \neq V_\mathcal{E}$. Should the sum of the resistor voltages be equal to the battery voltage? Consider a similar question: in a large commercial building there are four stairs that go from the first floor to the second floor, so is the second floor 4 stories above the first floor? No, because each person only takes one of the stairs, not all of them, in going from the first to the second floor. Likewise each charge goes through either the 10 Ω resistor or the 15 Ω resistor but not both as it moves through the circuit.

EXAMPLE

What is the voltage across the 6 Ω resistor?

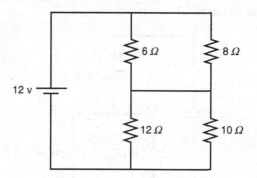

Is the 6 Ω resistor in series with the 12 Ω resistor? Current that passes through the 6 Ω resistor could go through the 12 Ω resistor but it could go to the right and through either the 8 Ω resistor or the 10 Ω resistor, so the 6 Ω and 12 Ω resistors are not in series.

Use the finger test with the 6 Ω resistor: start with a finger at each end and move them along only wires until they are at the two ends of something else. The 8 Ω resistor is in parallel with the 6 Ω resistor. The resistance of the top pair is

$$\frac{1}{R_{\text{top}}} = \frac{1}{6\,\Omega} + \frac{1}{8\,\Omega} = \frac{8\,\Omega}{48\,(\Omega)^2} + \frac{6\,\Omega}{48\,(\Omega)^2}$$

$$R_{\text{top}} = \frac{48\,(\Omega)^2}{14\,\Omega} = 3.43\,\Omega$$

Likewise the 12 Ω resistor and the 10 Ω resistor are in parallel.

$$\frac{1}{R_{\text{bottom}}} = \frac{1}{12\,\Omega} + \frac{1}{10\,\Omega} = \frac{10\,\Omega}{120\,(\Omega)^2} + \frac{12\,\Omega}{120\,(\Omega)^2}$$

$$R_{\text{bottom}} = \frac{120\,(\Omega)^2}{22\,\Omega} = 5.45\,\Omega$$

The top pair of resistors is in series. The current leaves the battery, goes through the top pair of resistors (now replaced with a single resistor), then goes through the bottom pair of resistors (one way or the other), and back into the battery.

$$R_{\text{all}} = R_{\text{top}} + R_{\text{bottom}} = 3.43 \ \Omega + 5.45 \ \Omega = 8.88 \ \Omega$$

$$I_{\mathcal{E}} = I_{\text{all}} = \frac{V_{\text{all}}}{R_{\text{all}}} = \frac{12 \text{ V}}{8.88 \ \Omega} = 1.35 \text{ A}$$

Not all of this current passes through the 6 Ω resistor; it passes through the top pair.

$$V_6 = V_{\text{top}} = I_{\text{top}} R_{\text{top}} = (1.35 \text{ A})(3.43 \ \Omega) = 4.6 \text{ V}$$

The voltage on the 8 Ω resistor is also 4.6 V.

EXAMPLE

What is the charge on the 10 μF capacitor?

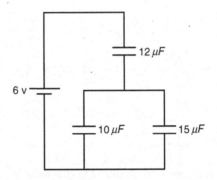

The 10 μF capacitor is not in parallel with the battery, so the voltage across it isn't 6 V. Nor is the 10 μF capacitor in series with the 12 μF capacitor. Charge that passes through the 12 μF capacitor could go through the 10 μF capacitor but it could go through the 15 μF capacitor, so the 10 μF and 12 μF capacitors are not in series.

Is the 10 μF capacitor in parallel with the 15 μF capacitor? Do the finger test, starting at the two ends of the 10 μF capacitor. You can move the fingers to the two ends of the 15 μF capacitor by moving only along wires, without crossing any circuit elements. They are in parallel, and we can replace them with a single capacitor.

$$C_{\text{pair}} = C_{10} + C_{15} = 10 \ \mu\text{F} + 15 \ \mu\text{F} = 25 \ \mu\text{F}$$

This pair is in series with the 12 μF capacitor. Current that goes through the 12 μF capacitor must go through the pair, passing through either the 10 μF capacitor or the 15 μF capacitor. The 12 μF capacitor and the 10/15 pair can be replaced with a single capacitor.

$$\frac{1}{C_{\text{trio}}} = \frac{1}{12 \ \mu\text{F}} + \frac{1}{25 \ \mu\text{F}} = \frac{25 \ \mu\text{F}}{300 \ (\mu\text{F})^2} + \frac{12 \ \mu\text{F}}{300 \ (\mu\text{F})^2}$$

$$C_{\text{trio}} = \frac{300 \ (\mu\text{F})^2}{37 \ \mu\text{F}} = 8.1 \ \mu\text{F}$$

The charge on the 8.1 μF capacitor trio is

$$Q_{trio} = C_{trio}V_{trio} = (8.1\ \mu F)(6\ V) = 48.6\ \mu C$$

This charge goes through both the 12 μF capacitor and the 25 μF capacitor pair, so the charge on the 12 μF capacitor is 48.6 μC and the charge on the 25 μF capacitor pair is 48.6 μC.

The voltage across the pair is

$$Q_{pair} = C_{pair}V_{pair}$$

$$48.6\ \mu C = (25\ \mu F)V_{pair}$$

$$V_{pair} = 1.94\ V$$

Because the 10 μF capacitor and the 15 μF capacitor are in parallel,

$$V_{10} = V_{15} = V_{pair} = 1.94\ V$$

The charge on the 10 μF capacitor is

$$Q_{10} = C_{10}V_{10} = (10\ \mu F)(1.94\ V) = 19.4\ \mu C$$

LESSON

Look for series and parallel combinations. Work your way out until you find a combination that you know the voltage across.

EXAMPLE

What is the current through the 5 Ω resistor in each circuit?

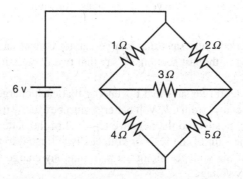

There are no two resistors in the first circuit that are in series or in parallel, so we can't combine resistors. Neither the 5 Ω resistor nor any other resistor is in parallel with the battery, so we don't know the voltage across any of the resistors. Since we can't reduce the circuit to something simpler we can't solve this circuit (yet).

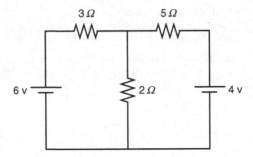

For the second circuit, current that goes through the 5 Ω resistor could go through the 2 Ω resistor but it could go through the 3 Ω resistor, so the 5 Ω resistor and the 2 Ω resistor are not in series. None of the resistors are in series or in parallel, so we can't simplify the circuit. (The 5 Ω resistor is in series with the 4 V battery, but we don't have a replacement for such a pair.)

The voltage on the 5 Ω resistor is not 4 V because the 5 Ω resistor is not in parallel with the 4 V battery—we don't know the voltage across the 5 Ω resistor.

Both of these circuits are too complex to solve with only our series and parallel technique.

11.4 KIRCHHOFF'S RULES

Kirchhoff has two rules for dealing with circuits:

- The **node rule**: The current into any point or element in the circuit is the same as the current out of that point or element.
- The **loop rule**: The sum of the voltage changes along any closed path in the circuit is zero.

Kirchhoff's node rule says that charge cannot collect anywhere. This is the same idea we used in the last sections to say that two devices in series had to have the same current.

Kirchhoff's loop rule says that the energy that each charge has (12 joules for each coulomb if it comes from a 12 volt battery) must be exactly used up by the charge as it goes from the positive to the negative terminal of the battery. This is why in a complex circuit the sum of the resistor voltages is not equal to the battery voltage. Instead the sum of the voltages along the path that any charge takes adds up to be equal to the battery voltage.

EXAMPLE

Find the current through and voltage across the 4 Ω resistor.

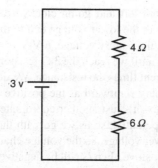

Starting at the bottom left and going clockwise, we first encounter the battery. We are moving from the lower voltage to the higher voltage end, so we are going up in voltage by the battery voltage, 3 V.

We continue around until we reach the 4 Ω resistor. What is the voltage change as we cross the 4 Ω resistor? The voltage across the 4 Ω resistor is equal to the current times the resistance. We don't know the current, so we give it a symbol i, choosing downward as the positive direction. Current goes from higher potential to lower potential, so as we go with the current we go "downhill," from higher voltage to lower voltage, so the voltage change is $-(4\ \Omega)i$.

Next we get to the 6 Ω resistor. The voltage across the 6 Ω resistor is equal to the current times the resistance. The two resistors are in series, so the current through the 6 Ω resistor is also i downward. The voltage change is $-(6\ \Omega)i$.

Now we're back where we started, so the loop rule says that the sum of the voltage changes is zero.

$$+3\ \text{V} - (4\ \Omega)i - (6\ \Omega)i = 0$$

$$+3\ \text{V} - (4\ \Omega + 6\ \Omega)i = 0$$

$$i = \frac{3\ \text{V}}{(4\ \Omega + 6\ \Omega)} = 0.30\ \text{A}$$

The voltage across the 4 Ω resistor is the resistance times the current through the resistor.

$$V_4 = I_4 R_4 = (0.30\ \text{A})(4\ \Omega) = 1.2\ \text{V}$$

EXAMPLE

What is the current through the 5 Ω resistor?

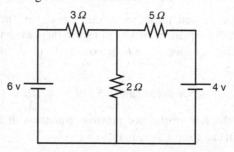

Starting at the bottom left and going clockwise, we first encounter the 6V battery. We are moving from the lower voltage end to the higher voltage end, so we are going up in voltage by the battery voltage, 6 V.

We continue around until we reach the 3 Ω resistor. The voltage across the 3 Ω resistor is equal to the current times the resistance. We don't know the current, so we give it a symbol i_3, choosing rightward as the positive direction. We could choose either direction as positive—it's just another coordinate system. Current goes from higher potential to lower potential, so as we go with the current we go "downhill," from higher voltage to lower voltage, so the voltage change is $-(3\ \Omega)i_3$. If we chose the "wrong" direction as positive, then i_3 will be negative and the voltage change will be positive.

Next we encounter the 2 Ω resistor. Current that goes through the 3 Ω resistor could go through the 2 Ω resistor or it could go through the 5 Ω resistor, so the 3 Ω resistor and the 2 Ω resistor are not in series. The current through the 2 Ω resistor is not necessarily the same as the current through the 3 Ω resistor, so we need a new symbol i_2 to represent the current through the 2 Ω resistor. We choose downward as positive i_2 and the voltage change is $-(2\ \Omega)i_2$.

Continuing in a clockwise fashion, we get back to where we started. The sum of the voltage changes must be zero.

$$+ 6\text{V} - (3\ \Omega)i_3 - (2\ \Omega)i_2 = 0$$

This equation has two unknowns, so we can't solve it yet. We need another equation.

We go around another loop, starting at the middle of the bottom and going up and clockwise. First we encounter the 2 Ω resistor, but this time we're going against the current (we've already chosen the positive direction and we need to stick with our choice). Current flows from higher potential to lower potential, so we're moving from lower potential to higher potential, or up in voltage. The voltage change across the 2 Ω resistor is $+(2\ \Omega)i_2$.

We turn right across the 5 Ω resistor. The current through the 5 Ω resistor is not equal to the current through either of the other resistors, so we need a new symbol i_5. We choose rightward as positive and the voltage change is $-(5\ \Omega)i_5$.

Last we cross the 4 V battery. We are moving from the higher voltage side to the lower voltage side, so the voltage change is -4 V. We're back where we started, so the sum of the voltage changes is zero.

$$+(2\ \Omega)i_2 - (5\ \Omega)i_5 - 4\text{ V} = 0$$

This equation also has two unknowns, so we can't solve it yet. The unknowns aren't the same two as in the other equation, so we can't solve them as simultaneous equations.

We need another equation. If we went around the outside of the circuit we would get

$$+ 6\text{ V} - (3\ \Omega)i_3 - (5\ \Omega)i_5 - 4\text{ V} = 0$$

which is equal to the sum of the two previous equations. It isn't an independent equation, so we still can't solve the equations.

Instead we apply the node rule. Looking at the middle of the top of the circuit, the sum of the currents in is equal to the sum of the currents out:

$$i_3 = i_2 + i_5$$

This gives us three equations and three unknowns

$$\begin{cases} +6\text{ V} - (3\ \Omega)i_3 - (2\ \Omega)i_2 = 0 \\ + (2\ \Omega)i_2 - (5\ \Omega)i_5 - 4\text{ V} = 0 \\ \qquad i_3 = i_2 + i_5 \end{cases}$$

$$+6\text{ V} - (3\ \Omega)(i_2 + i_5) - (2\ \Omega)i_2 = 0$$

$$+6\text{ V} - (3\ \Omega)i_5 - (5\ \Omega)i_2 = 0$$

$$+(5\ \Omega)i_2 = +6\text{ V} - (3\ \Omega)i_5$$

$$+(2\ \Omega)\left(\frac{+6\text{ V} - (3\ \Omega)i_5}{+(5\ \Omega)}\right) - (5\ \Omega)i_5 - 4\text{ V} = 0$$

$$+2.4\text{ V} - (1.2\ \Omega)i_5 - (5\ \Omega)i_5 - 4\text{ V} = 0$$

$$-(6.2\ \Omega)i_5 = 1.6\text{ V}$$

$$i_5 = -0.26\text{ A}$$

The minus sign indicates that the current through the 5 Ω resistor goes in the negative direction, opposite the direction we chose as positive, or leftward.

EXAMPLE

What is the current through the 12 Ω resistor?

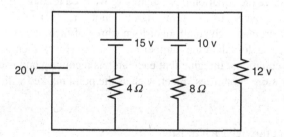

We could go around each of the three smallest loops:

$$\begin{cases} +20\text{ V} - 15\text{ V} - (4\ \Omega)i_4 = 0 \\ + (4\ \Omega)i_4 + 15\text{ V} - 10\text{ V} - (8\ \Omega)i_8 = 0 \\ + (8\ \Omega)i_8 + 10\text{ V} - (12\ \Omega)i_{12} = 0 \end{cases}$$

We could solve these equations to find i_{12}.

Alternatively, we could go around the outside of the circuit:

$$+20 \text{ V} - (12 \ \Omega)i_{12} = 0$$

This we can solve immediately.

$$-(12 \ \Omega)i_{12} = -20 \text{ V}$$

$$i_{12} = 1.67 \text{ A}$$

LESSON

Look for loops with the fewest number of variables—solving the equations might be easy.

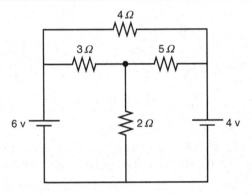

Sometimes it can be difficult to get a matching number of variables and independent equations. Consider the circuit above. If the goal is to find the current through the 4 Ω resistor then we can go around the outside, along a loop with no other resistors, and solve the equation immediately. If we need to find something else, then we'll need a complete set of equations. If you're not sure when you've got a complete set, do this: go around each small inner loop (three of them here). If the number of equations (three) is less than the number of variables (four), then use the node rule to get enough equations. Imagine that each area is a country and use the node rules where three or more "countries" meet, such as the point marked with the gray circle.

EXAMPLE

Find the current through each resistor.

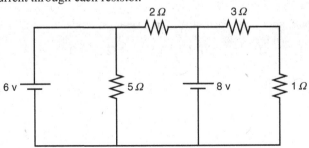

Do the finger test on the 5 Ω resistor, and you can see that it is in parallel with the 12 V battery. The voltage across the 5 Ω resistor is the same as the voltage across the 12 V battery, or 12 V.

$$i_5 = \frac{V_5}{R_5} = \frac{12\ \text{V}}{5\ \Omega} = 2.4\ \text{A}$$

The current goes from higher battery voltage to lower battery voltage, or downward.

For the 2 Ω resistor, we look for a loop with the fewest resistors possible. We find one using the two batteries, picking rightward as positive.

$$+12\ \text{V} - (2\ \Omega)i_2 - 8\ \text{V} = 0$$

$$-(2\ \Omega)i_2 = -4\ \text{V}$$

$$i_2 = +2\ \text{A}$$

The 3 Ω resistor is in series with the 1 Ω resistor, so they act like a single 4 Ω resistor. This 4 Ω pair is in parallel with the 8 V battery, so

$$i_3 = i_1 = i_{\text{pair}} = \frac{8\ \text{V}}{4\ \Omega} = 2\ \text{A}$$

from the higher voltage terminal of the battery to the lower voltage terminal.

11.5 MIXED CIRCUITS: R-C AND R-L

There are two more basic circuit elements (so far we have battery, wire, resistor, and capacitor). A switch is a device for shutting off current, so a "closed" switch looks like a complete circuit and an "open" switch looks like an incomplete circuit. An inductor is a coil or solenoid and is represented by curved half-loops.

The voltage across a resistor obeys Ohm's law: the voltage is proportional to the current at the time. The voltage across a capacitor depends on the charge, which is the sum of all of the current that has flowed through the capacitor:

$$V_C = \frac{Q}{C} = \frac{1}{C} \int i(t)\, dt$$

The voltage across an inductor depends on how the current through the inductor is changing—in a way, it depends on the current through the inductor in the future

$$V_L = L \frac{di}{dt}$$

where L is the inductance in henrys (H). Some physics texts have a minus sign in this equation so that it matches the equation for induced voltage, but that is incorrect for analyzing circuits.

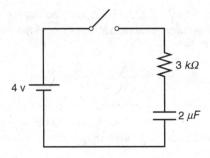

Consider what happens when we mix a resistor and a capacitor in a circuit. At the instant that we close the switch (shown open), the charge on the capacitor is zero—there hasn't been time for any current to flow through it yet. Since the charge is zero, the voltage across the capacitor is zero. The loop rule tells us that the voltage across the resistor is 4 V, so the current in the circuit is

$$I = \frac{V}{R} = \frac{4 \text{ V}}{3 \text{ k}\Omega} = 1.3 \text{ mA}$$

The capacitor is charging at 1.3 mC each second. As the charge on the capacitor increases, so does the voltage across the capacitor, and the voltage across the resistor decreases, so the current deceases. Eventually the voltage on the capacitor reaches 4 V and the current stops.

If we use the loop rule, we get

$$(4 \text{ V}) - (3 \text{ k}\Omega)i - \frac{1}{(2 \text{ }\mu\text{F})} \int_0^t i(t') \, \mathrm{d}t' = 0$$

where t is the time at which we want the current, but we've already used it so we can't use it as the variable of integration, so we pick another symbol t'. The solution to the differential equation is

$$i(t) = (1.3 \text{ mA}) \, e^{-t/\tau} \quad \text{where} \quad \tau = RC = (3 \text{ k}\Omega)(2 \text{ }\mu\text{F}) = 6 \text{ ms}$$

Whenever we mix a capacitor or an inductor with a resistor we get a current that depends on time in an exponential fashion. I will not cover how to solve these equations in this book. Instead, we will discuss how to handle the extremes: just before and just after the switch has been closed.

For the charge on a capacitor to change, a current must flow through the capacitor. For the charge to change instantaneously, an infinite current must flow through the capacitor. Since we can't have an infinite current, the charge on a capacitor can't change instantaneously, so the voltage across a capacitor can't change instantaneously. The voltage can change, and we can measure the instantaneous rate of change (as opposed to the average rate of change), but there can't be a discrete jump, such as where the voltage is 2 V and then an instant later is 4 V without going through 3 V.

For an inductor, if the current were to change instantaneously, then the voltage would have to be infinite, which can't happen. If you try to switch off a circuit with an inductor in it, stopping the current causes such a high voltage across the switch that the charge leaps across the switch, causing a spark.

On the other hand, if we wait for things to settle to a steady state situation, then nothing must be changing. If there were to be a current through any capacitor, then the charge would be changing and the voltage would be changing. If there were to be a voltage on an inductor, then the current would be changing. So after a long time, the voltage across any inductor must be zero and the current through any capacitor must be zero.

EXAMPLE

The switch is connected to the left so that the capacitor can charge completely. The switch is then thrown to the right. What is the current through each resistor immediately after the switch is thrown to the right?

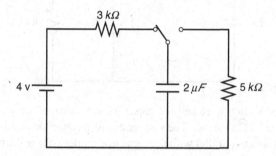

Once the switch is thrown to the right, the 3 kΩ resistor is no longer part of a complete circuit—there is no closed path around the circuit that includes the 3 kΩ resistor. Therefore the current through the 3 kΩ resistor must be zero.

After the switch is thrown to the right, the 5 kΩ resistor is in parallel with the capacitor—apply the finger test. The voltage across the 5 kΩ resistor must be the same as the voltage across the capacitor. The current through the 5 kΩ resistor is the capacitor voltage divided by the resistance.

The voltage across a capacitor cannot change instantaneously; it takes time for charge to flow. So the voltage across the capacitor just after the switch is thrown to the right is the same as the voltage just before it was thrown. At that time it was in a circuit with the battery and the 3 kΩ resistor. Since that circuit had been connected for a long time ("charge completely"), the current in the left circuit was zero just before the switch was thrown. Since there was no current, there was no voltage on the resistor, and the capacitor voltage was equal to the battery voltage

$$+(4 \text{ V}) - (3 \text{ k}\Omega)(0 \text{ A}) - V_C = 0$$

$$V_C = 4 \text{ V}$$

$$I_{5k} = \frac{V_{5k}}{R_{5k}} = \frac{V_C}{R_{5k}} = \frac{4 \text{ V}}{5 \text{ k}\Omega} = 0.8 \text{ mA}$$

This value does not depend on the capacitance of the capacitor. However, if a larger capacitor is used, then it will take longer for the capacitor to discharge through the 5 kΩ resistor. In this circuit, the charge will be 99% discharged in 50 ms.

EXAMPLE

What is the current through the 16 Ω resistor (a) immediately after the switch is closed, (b) a long time after the switch is closed, and (c) immediately after the switch is then opened?

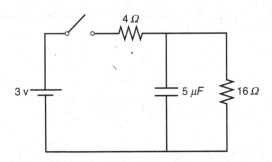

Before the switch is closed the capacitor is uncharged (any charge will have gone through the 16 Ω resistor). The voltage across a capacitor with no charge is zero. The 16 Ω resistor is in parallel with the capacitor, so the voltages across them are the same, and the voltage across the 16 Ω resistor is zero.

Immediately after the switch is closed, the charge on the capacitor is still zero—it hasn't had time to change because there hasn't been time for charge to flow yet. Since the capacitor is uncharged and has zero volts across it, there is zero volts across the 16 Ω resistor. There is 0.75 A of current flowing through the 4 Ω resistor, but since there is none in the 16 Ω resistor all of the current goes through the capacitor, charging it.

After a long time has passed, the capacitor is fully charged. If there was a current through the capacitor then its charge would be changing, so there is no current through the capacitor. All of the current through the 4 Ω resistor goes through the 16 Ω resistor–they are effectively in series now (only when the capacitor is fully charged). There is a current of

$$I = \frac{V}{R} = \frac{3 \text{ V}}{4\,\Omega + 16\,\Omega} = 0.15 \text{ A}$$

through the two resistors.

The switch is now opened. The voltage across the capacitor cannot change instantaneously, so it is the same as it was after the switch had been closed for a long time. The voltage across the capacitor is equal to the voltage across the 16 Ω resistor, so it was

$$V_C = V_{16} = I_{16}R_{16} = (0.15 \text{ A})(16\,\Omega) = 2.4 \text{ V}$$

After the switch is opened, the voltage on the capacitor and the 16 Ω resistor is still 2.4 V, so the current through the 16 Ω resistor is still 0.15 A. This will decrease exponentially to zero with a time constant ($1/e$ time) of 80 μs, so that the current will decrease to only 1% of its initial value in 0.4 ms.

EXAMPLE

What is the current through the 16 Ω resistor (a) immediately after the switch is closed, (b) a long time after the switch is closed, and (c) immediately after the switch is then opened?

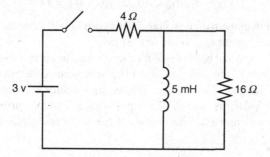

Before the switch is closed there is no current anywhere in the circuit. Immediately after the switch is closed, the current through the inductor is still zero—it hasn't had time to change yet. Since the current through the inductor is zero, all of the current through the 4 Ω resistor goes through the 16 Ω resistor. They are effectively in series now (only when the current through the inductor is zero). There is a current of

$$I = \frac{V}{R} = \frac{3\text{ V}}{4\,\Omega + 16\,\Omega} = 0.15\text{ A}$$

through the two resistors.

Though the current through the inductor is zero, the current is changing. The rate at which this current changes is

$$L\frac{di}{dt} = V_L = V_{16} = I_{16}R_{16}$$

$$(5\text{ mH})\frac{di}{dt} = (0.15\text{ A})(16\,\Omega) = 2.4\text{ V}$$

$$\frac{di}{dt} = \frac{2.4\text{ V}}{5\text{ mH}} = 480\text{ V/H} = 480\text{ A/s}$$

As more current goes through the inductor, less goes through the 16 Ω resistor, so the voltage across the resistor is less, so the voltage across the inductor is less, and the rate of change decreases. Eventually the current through the inductor reaches a steady-state value and doesn't change any more. Since the inductor current isn't changing, the voltage across the inductor is zero, so the voltage across the 16 Ω resistor is zero, so the current through the 16 Ω resistor is zero. There is still a current of

$$I = \frac{V}{R} = \frac{3\text{ V}}{4\ \Omega} = 0.75\text{ A}$$

going through the 4 Ω resistor and the inductor.

The switch is now opened. The current through the inductor cannot change instantaneously, so it is the same as it was after the switch had been closed for a long time, or 0.75 A. This current cannot go through the 4 Ω resistor, but only through the 16 Ω resistor. The current through the 16 Ω resistor immediately after the switch is opened is 0.75 A, and the voltage across the 16 Ω resistor is

$$V_{16} = I_{16}R_{16} = (0.75\text{ A})(16\ \Omega) = 12\text{ V}$$

which is greater than the battery voltage! This current will decrease exponentially to zero with a time constant ($1/e$ time) of 0.31 ms.

After the switch had been closed for a long time, the current through the inductor was downward. Switching directions would be an instantaneous change in the current, which can only happen if the voltage across the inductor is infinite. Therefore the current after the switch is opened is going downward, so the current going through the 16 Ω resistor is upward. The bottom of the 16 Ω resistor is 12 V higher than the top.

EXAMPLE

What is the current through the 6 Ω resistor (a) immediately after the switch is closed and (b) a long time after the switch is closed?

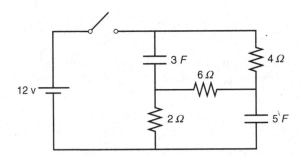

Before the switch is closed the capacitors are uncharged and the voltage across each capacitor is zero. Immediately after the switch is closed, the charge on the capacitors is still zero—it hasn't had time to change because there hasn't been time for charge to flow yet. The capacitors and the switch act (momentarily) like wires: no matter how much current is flowing through them the voltage across them is zero. Try the finger test on the 6 Ω resistor. Since you can pass through the capacitors and the switch you should find that the 6 Ω resistor is (momentarily) in parallel with the battery. The voltage across the 6 Ω resistor is 12 V and the current through it (left to right) is

$$I = \frac{V}{R} = \frac{12\ V}{6\ \Omega} = 2\ A$$

There will be a current of 8 A flowing through the 3 F capacitor and a current of 5 A flowing through the 5 F capacitor.

After a long time has passed, the capacitors will be fully charged. If there was a current through either capacitor then its charge would be changing, so there is no current through the capacitors.

All of the current through the 4 Ω resistor goes through the 6 Ω resistor and then the 2 Ω resistor. The resistors are effectively in series now (only when the capacitors are fully charged). There is a current of

$$I = \frac{V}{R} = \frac{12\ V}{4\ \Omega + 6\ \Omega + 2\ \Omega} = 1\ A$$

through the resistors, flowing right to left through the 6 Ω resistor.

CHAPTER SUMMARY

- The equations for resistors and capacitors can be used for individual circuit elements or for groups of similar elements.

$$V = IR \quad \text{(Ohm's law)}$$
$$P = IV$$
$$R_S = R_1 + R_2 \quad \text{and} \quad \frac{1}{R_P} = \frac{1}{R_1} + \frac{1}{R_2}$$
$$Q = CV$$
$$U_{CAP} = \frac{1}{2}QV$$
$$\frac{1}{C_S} = \frac{1}{C_1} + \frac{1}{C_2} \quad \text{and} \quad C_P = C_1 + C_2$$
$$V_L = L\frac{di}{dt}$$

- The **node rule**: The currents into and out of any point or element in the circuit are the same.
- The **loop rule**: The sum of the voltage changes along any closed path in the circuit is zero.
- Look for series and parallel combinations to simplify the circuit.
- Two items in series have the same current.
- Two items in parallel have the same voltage.

INDEX